Renewable

How Clean Power is Changing Our World

Lillian Benson

Renewable Energy Revolution

© *Copyright 2024 by **Lillian Benson***

All rights reserved

This document is geared towards providing exact and reliable information with regards to the topic and issue covered. The publication is sold with the idea that the publisher is not required to render accounting, officially permitted, or otherwise, qualified services. If advice is necessary, legal or professional, a practiced individual in the profession should be ordered.

From a Declaration of Principles which was accepted and approved equally by a Committee of the American Bar Association and a Committee of Publishers and Associations.

In no way is it legal to reproduce, duplicate, or transmit any part of this document in either electronic means or in printed format. Recording of this publication is strictly prohibited and any storage of this document is not allowed unless with written permission from the publisher. All rights reserved.

The information provided herein is stated to be truthful and consistent, in that any liability, in terms of inattention or otherwise, by any usage or abuse of any policies, processes, or directions contained within is the solitary and utter responsibility of the recipient reader. Under no circumstances will any legal responsibility or blame be held against the publisher for any reparation, damages, or monetary loss due to the information herein, either directly or indirectly.

Respective authors own all copyrights not held by the publisher.

The information herein is offered for informational purposes solely, and is universal as so. The presentation of the information is without contract or any type of guarantee assurance.

The trademarks that are used are without any consent, and the publication of the trademark is without permission or backing by the trademark owner. All trademarks and brands within this book are for clarifying purposes only and are owned by the owners themselves, not affiliated with this document.

TABLE OF CONTENTS

Chapter 1: Solar Energy: Harnessing the Power of the Sun 5

Photovoltaic Technology Explained 5

Solar Farms and Urban Integration 9

Innovations in Solar Panel Efficiency 13

Economic and Environmental Benefits 17

Overcoming Barriers to Adoption 21

Chapter 2: Wind Energy: Capturing Nature's Breeze 26

The Mechanics of Wind Turbines 26

Offshore vs. Onshore Wind Farms 30

Technological Advancements in Wind Power 33

Environmental and Economic Impacts 37

Policy and Regulatory Frameworks 41

Chapter 3: Hydropower: The Force of Water 46

Understanding Hydroelectric Systems 46

Large-Scale Dams vs. Small Hydro Projects 50

Environmental Considerations and Mitigation 53

The Role of Hydropower in Energy Grids 57

Future Trends and Innovations 60

Chapter 4: Biomass and Bioenergy: Organic Solutions 64

Types of Biomass and Conversion Processes 64

Biofuels: From Production to Consumption 68

Environmental and Economic Implications 72

Challenges in Scaling Bioenergy 76

Innovations and Future Directions 80

Chapter 5: Geothermal Energy: Tapping Earth's Heat 84

Geothermal Systems and Technologies 84

Global Geothermal Hotspots 88

Environmental and Economic Benefits 92

Chapter 1: Solar Energy: Harnessing the Power of the Sun

Photovoltaic Technology Explained

Photovoltaic technology, often referred to as solar power, represents a remarkable leap in harnessing the sun's energy to generate electricity. At its core, this technology relies on the photovoltaic effect, a process by which solar cells convert sunlight directly into electricity. These cells, typically made from silicon, are the building blocks of solar panels, which are becoming increasingly common on rooftops and in solar farms worldwide.

The journey of photovoltaic technology begins with the discovery of the photovoltaic effect in the 19th century. However, it wasn't until the latter half of the 20th century that significant advancements were made, leading to the development of the first practical solar cells. These early cells were primarily used in space applications, powering satellites where traditional energy sources were impractical. Over time, improvements in efficiency and reductions in cost have made photovoltaic technology accessible for terrestrial applications, transforming it into a viable option for renewable energy generation.

The basic principle behind photovoltaic technology involves the absorption of photons, the fundamental particles of light, by semiconductor materials within the solar cells. When these photons strike the surface of the cell, they transfer their energy to electrons, freeing them from their atomic bonds and allowing them to flow through the material as an electric current. This current is then captured by conductive metal plates on the sides of the cell and transferred to wires, forming an electrical circuit.

One of the most significant challenges in photovoltaic technology is improving the efficiency of solar cells. Efficiency refers to the percentage of sunlight that can be converted into usable electricity. Traditional silicon-based solar cells have an efficiency range of 15-20%, but recent advancements have pushed this figure higher. Researchers are exploring various materials and technologies, such as perovskite solar cells and multi-junction cells, which have the potential to achieve efficiencies exceeding 40%.

Perovskite solar cells, in particular, have garnered significant attention due to their high efficiency and relatively low production costs. These cells use a different type of semiconductor material that can be manufactured using simpler processes compared to traditional silicon cells. However, challenges remain in terms of stability and scalability, as perovskite materials can degrade over time when exposed to environmental factors.

Another promising development in photovoltaic technology is the advent of bifacial solar panels. Unlike traditional panels that capture sunlight only from one side, bifacial panels can absorb light from both sides, increasing their overall energy output. This is particularly advantageous in environments with reflective surfaces, such as snow-covered areas or bodies of water, where sunlight can be reflected onto the rear side of the panels.

The integration of photovoltaic technology into urban environments presents unique opportunities and challenges. Rooftop solar installations are becoming increasingly popular in residential and commercial buildings, allowing individuals and businesses to generate their own electricity and reduce reliance on the grid. However, urban integration requires careful consideration of factors such as shading from nearby structures, roof orientation, and available space.

Innovations in solar panel design are addressing these challenges by creating flexible and lightweight panels that can be installed on a variety of surfaces, including curved or irregularly shaped roofs. Building-integrated photovoltaics (BIPV) represent another exciting development, where solar cells are seamlessly integrated into building materials such as windows, facades, and roof tiles. This approach not only generates electricity but also enhances the aesthetic appeal of buildings.

The economic and environmental benefits of photovoltaic technology are substantial. On the economic front, the cost of solar panels has plummeted over the past decade, making solar power one of the most cost-effective sources of electricity in many regions. This cost reduction is largely attributed to advancements in manufacturing processes, economies of scale, and increased competition in the solar industry. As a result, solar power is becoming increasingly competitive with traditional fossil fuels, driving its adoption worldwide.

From an environmental perspective, photovoltaic technology offers a clean and sustainable energy solution. Solar power generation produces no greenhouse gas emissions, air pollutants, or water pollution, making it an environmentally friendly alternative to fossil fuels. Additionally, solar panels have a relatively low environmental impact during their production and can be recycled at the end of their lifespan, further enhancing their sustainability.

Despite these advantages, there are barriers to the widespread adoption of photovoltaic technology. One of the primary challenges is the intermittent nature of solar power, as electricity generation is dependent on sunlight availability. This variability necessitates the development of energy storage solutions, such as batteries, to store excess energy generated during sunny periods for use during cloudy days or at night.

Grid integration is another challenge, as the existing electrical grid infrastructure was not designed to accommodate the decentralized nature of solar power generation. Upgrading grid systems to handle distributed energy sources and implementing smart grid technologies are essential steps in overcoming this barrier.

Policy and regulatory frameworks play a crucial role in promoting the adoption of photovoltaic technology. Government incentives, such as tax credits, rebates, and feed-in tariffs, have been instrumental in driving the growth of the solar industry. However, these policies must be carefully designed to ensure long-term sustainability and avoid market distortions.

Public awareness and education are also vital in overcoming barriers to adoption. Many individuals and businesses remain unaware of the benefits and potential of photovoltaic technology, leading to misconceptions and hesitancy in embracing solar power. Educational campaigns and outreach programs can help dispel myths and provide accurate information about the advantages of solar energy.

In conclusion, photovoltaic technology represents a transformative force in the global energy landscape. Its ability to harness the sun's abundant energy offers a sustainable and environmentally friendly solution to the world's growing energy demands. As advancements continue to improve efficiency, reduce costs, and address integration challenges, photovoltaic technology is poised to play a pivotal role in the transition to a renewable energy future.

Solar Farms and Urban Integration

Solar farms, vast expanses of photovoltaic panels, have become emblematic of the renewable energy revolution. These installations, often located in rural or semi-rural areas, harness the sun's energy on a large scale, feeding electricity into the grid and reducing reliance on fossil fuels. The concept of solar farms is straightforward: by covering large areas with solar panels, they capture sunlight and convert it into electricity, providing a clean and sustainable energy source. However, the integration of solar technology into urban environments presents a unique set of challenges and opportunities, requiring innovative approaches to maximize efficiency and minimize disruption.

The development of solar farms has been driven by the need to generate significant amounts of renewable energy to meet growing demands. These installations can range from a few acres to several square miles, depending on the available land and the energy requirements of the region. The choice of location is critical, as solar farms require unobstructed sunlight and minimal shading from nearby structures or vegetation. Additionally, proximity to existing electrical infrastructure is essential to facilitate the efficient transmission of electricity to the grid.

One of the primary advantages of solar farms is their ability to generate large quantities of electricity without producing greenhouse gas emissions or air pollutants. This makes them an attractive option for regions seeking to reduce their carbon footprint and transition to cleaner energy sources. Furthermore, solar farms can be constructed relatively quickly compared to traditional power plants, allowing for rapid deployment in response to energy shortages or policy mandates.

Despite these benefits, solar farms face several challenges, particularly in terms of land use and environmental impact. The construction of large-scale solar installations requires significant land area, which can lead to conflicts with agricultural activities, wildlife habitats, and local communities. To address these concerns, developers are increasingly exploring dual-use strategies, where solar panels are installed alongside agricultural activities, allowing for the simultaneous production of food and energy. This approach, known as agrivoltaics, can enhance land productivity and provide additional income streams for farmers.

Urban integration of solar technology presents a different set of challenges and opportunities. Cities, with their dense populations and high energy demands, offer significant potential for solar energy generation. However, the limited availability of open space and the presence of tall buildings can complicate the installation of solar panels. To overcome these obstacles, urban planners and architects are exploring innovative solutions that incorporate solar technology into the fabric of the city.

Rooftop solar installations are one of the most common methods of integrating solar technology into urban environments. By utilizing existing roof space, these systems can generate electricity without requiring additional land. This approach is particularly effective in densely populated areas, where land is scarce and expensive. However, the success of rooftop solar installations depends on several factors, including roof orientation, structural integrity, and shading from nearby buildings or trees.

Building-integrated photovoltaics (BIPV) represent another promising avenue for urban solar integration. This approach involves incorporating solar cells directly into building materials, such as windows, facades, and roof tiles. BIPV systems offer several advantages, including aesthetic appeal, space efficiency, and the potential for energy savings through passive solar design. However, the widespread adoption of BIPV is hindered by higher costs and technical challenges related to integrating solar technology into traditional building materials.

Community solar projects offer an alternative model for urban solar integration, allowing residents and businesses to collectively invest in and benefit from solar energy. These projects typically involve the installation of a shared solar array, with participants receiving credits on their electricity bills based on their share of the generated power. Community solar projects can overcome barriers related to individual rooftop installations, such as shading or structural limitations, and provide access to solar energy for renters or those without suitable roof space.

The economic implications of urban solar integration are significant, with the potential to reduce energy costs, create jobs, and stimulate local economies. The declining cost of solar technology, combined with government incentives and favorable financing options, has made solar energy increasingly accessible to urban residents and businesses. Additionally, the growth of the solar industry has created a demand for skilled workers in areas such as installation, maintenance, and manufacturing, providing employment opportunities in communities transitioning to renewable energy.

Environmental benefits of urban solar integration extend beyond the reduction of greenhouse gas emissions. By generating electricity locally, solar installations can reduce the need for long-distance transmission, minimizing energy losses and enhancing grid resilience. Furthermore, the adoption of solar technology can contribute to improved air quality and public health by reducing reliance on fossil fuels and decreasing emissions of harmful pollutants.

Despite the numerous advantages of urban solar integration, several barriers remain. Regulatory and policy frameworks can pose challenges, as building codes, zoning regulations, and permitting processes may not be well-suited to accommodate solar installations. Streamlining these processes and providing clear guidelines for solar development can facilitate the adoption of solar technology in urban areas.

Public awareness and education are also critical to the success of urban solar integration. Many individuals and businesses remain unaware of the benefits and potential of solar energy, leading to misconceptions and hesitancy in embracing solar technology. Educational campaigns and outreach programs can help dispel myths and provide accurate information about the advantages of solar energy, encouraging broader adoption.

In summary, the integration of solar technology into urban environments offers a promising pathway to sustainable energy generation. By leveraging innovative approaches and overcoming existing barriers, cities can harness the power of the sun to meet their energy needs, reduce their carbon footprint, and enhance the quality of life for their residents. As technology continues to advance and costs decline, the potential for urban solar integration will only grow, paving the way for a cleaner and more sustainable future.

Innovations in Solar Panel Efficiency

Solar panel efficiency has been a focal point of research and development in the renewable energy sector, as improving the ability of panels to convert sunlight into electricity directly impacts the viability and cost-effectiveness of solar energy. The quest for higher efficiency has led to a variety of innovations, each contributing to the evolution of solar technology and its integration into our energy systems.

The traditional silicon-based solar cells, which have dominated the market for decades, typically achieve efficiencies between 15% and 20%. While this range has been sufficient to drive the initial adoption of solar technology, the push for greater efficiency has spurred the exploration of new materials and cell designs. One of the most promising developments in this area is the emergence of perovskite solar cells. These cells, named after the mineral with a similar crystal structure, have demonstrated remarkable efficiency gains in laboratory settings, with some reaching over 25%. The appeal of perovskite cells lies not only in their efficiency but also in their potential for low-cost production, as they can be manufactured using simpler processes compared to traditional silicon cells.

Despite their promise, perovskite solar cells face challenges related to stability and scalability. The materials used in these cells can degrade when exposed to moisture, oxygen, and heat, which poses a significant hurdle for their commercial deployment. Researchers are actively working to address these issues by developing protective coatings and exploring alternative materials that maintain the desirable properties of perovskites while enhancing their durability.

Another innovation in solar panel efficiency is the development of multi-junction solar cells. These cells are designed to capture a broader spectrum of sunlight by stacking multiple layers of semiconductor materials, each optimized to absorb different wavelengths of light. This approach allows multi-junction cells to achieve efficiencies exceeding 40%, making them particularly attractive for applications where space is limited, such as in satellites or high-performance solar concentrators. However, the complexity and cost of manufacturing multi-junction cells have limited their widespread adoption, prompting ongoing research to simplify production processes and reduce costs.

Bifacial solar panels represent another advancement in the quest for higher efficiency. Unlike traditional panels that capture sunlight only from one side, bifacial panels are designed to absorb light from both the front and rear surfaces. This design allows them to take advantage of reflected sunlight from the ground or surrounding surfaces, increasing their overall energy output. Bifacial panels are particularly effective in environments with high albedo, such as snowy regions or areas with reflective surfaces like water bodies. The increased energy yield from bifacial panels can offset their higher initial cost, making them an attractive option for large-scale solar installations.

The integration of advanced materials and technologies into solar panels has also led to the development of thin-film solar cells. These cells use layers of semiconductor materials that are only a few micrometers thick, allowing them to be flexible and lightweight. Thin-film solar cells can be manufactured using a variety of materials, including cadmium telluride (CdTe), copper indium gallium selenide (CIGS), and amorphous silicon. While their efficiency is generally lower than that of traditional silicon cells, thin-film cells offer advantages in terms of versatility and ease of installation, making them suitable for applications where traditional panels may not be feasible.

The pursuit of higher efficiency has also driven innovations in solar panel design and installation. For example, solar tracking systems, which adjust the orientation of panels to follow the sun's path across the sky, can significantly increase energy capture by maintaining optimal angles for sunlight absorption throughout the day. While tracking systems add complexity and cost to solar installations, the increased energy yield can justify the investment, particularly in large-scale solar farms.

In addition to technological advancements, improvements in manufacturing processes have played a crucial role in enhancing solar panel efficiency. Techniques such as passivated emitter and rear cell (PERC) technology, which involves adding a reflective layer to the rear side of solar cells, have been widely adopted to boost efficiency by reducing energy losses. Similarly, the use of anti-reflective coatings and textured surfaces can minimize the reflection of sunlight, ensuring that more photons are absorbed by the solar cells.

The economic implications of these innovations are significant, as higher efficiency translates to lower costs per watt of electricity generated. This reduction in cost has been a key driver of the rapid growth of the solar industry, making solar energy increasingly competitive with traditional fossil fuels. As efficiency continues to improve, the levelized cost of electricity (LCOE) from solar installations is expected to decline further, accelerating the transition to renewable energy sources.

Environmental benefits of increased solar panel efficiency are equally important. By generating more electricity from the same amount of sunlight, higher efficiency panels reduce the land and resources required for solar installations, minimizing their environmental footprint. Additionally, the use of advanced materials and manufacturing techniques can reduce the energy and emissions associated with panel production, further enhancing the sustainability of solar technology.

Despite the progress made in solar panel efficiency, challenges remain in terms of scalability and integration into existing energy systems. The intermittent nature of solar power, driven by variations in sunlight availability, necessitates the development of energy storage solutions and grid management strategies to ensure a reliable and stable energy supply. Advances in battery technology and smart grid systems are essential components of this effort, enabling the efficient storage and distribution of solar energy.

Public awareness and education are also critical to the successful adoption of high-efficiency solar technology. Many individuals and businesses remain unaware of the latest advancements and their potential benefits, leading to misconceptions and hesitancy in embracing solar energy. Educational campaigns and outreach programs can help dispel myths and provide accurate information, encouraging broader adoption and investment in solar technology.

The innovations in solar panel efficiency represent a pivotal chapter in the ongoing evolution of renewable energy. By harnessing cutting-edge materials and technologies, the solar industry is poised to deliver cleaner, more affordable, and more sustainable energy solutions. As research and development continue to push the boundaries of what is possible, the future of solar energy looks brighter than ever, offering a path toward a more sustainable and resilient energy landscape.

Economic and Environmental Benefits

The transition to renewable energy sources is not only a technological shift but also an economic and environmental imperative. Solar energy, in particular, offers a compelling case for its widespread adoption due to its dual benefits: economic viability and environmental sustainability. Understanding these benefits is crucial for individuals, businesses, and policymakers as they navigate the complexities of energy transition.

Economically, solar energy has undergone a remarkable transformation over the past few decades. The cost of solar panels has plummeted, driven by advancements in technology, economies of scale, and increased competition in the solar industry. This reduction in cost has made solar power one of the most affordable sources of electricity in many parts of the world. For homeowners and businesses, investing in solar panels can lead to significant savings on electricity bills. By generating their own electricity, they can reduce or even eliminate their reliance on the grid, shielding themselves from rising energy prices and market volatility.

The financial benefits of solar energy extend beyond direct savings on electricity bills. Many governments offer incentives to encourage the adoption of solar technology, such as tax credits, rebates, and feed-in tariffs. These incentives can significantly reduce the upfront cost of installing solar panels, making the investment more accessible to a broader range of consumers. Additionally, solar installations can increase property values, as homes and businesses with solar panels are often seen as more attractive to buyers due to their lower operating costs and environmental credentials.

For businesses, the economic benefits of solar energy are multifaceted. In addition to reducing operational costs, adopting solar technology can enhance a company's brand image and reputation. Consumers are increasingly conscious of environmental issues and are more likely to support businesses that demonstrate a commitment to sustainability. By investing in solar energy, companies can differentiate themselves from competitors and attract environmentally conscious customers. Moreover, businesses that generate their own electricity can benefit from increased energy security and resilience, reducing the risk of disruptions caused by power outages or supply chain issues.

The economic impact of solar energy extends to job creation and economic development. The growth of the solar industry has led to the creation of millions of jobs worldwide, spanning manufacturing, installation, maintenance, and research and development. These jobs are often local and cannot be outsourced, providing a boost to local economies and contributing to community development. As the demand for solar energy continues to rise, the industry is expected to generate even more employment opportunities, supporting a just transition to a low-carbon economy.

From an environmental perspective, the benefits of solar energy are equally compelling. Solar power generation produces no greenhouse gas emissions, air pollutants, or water pollution, making it a clean and sustainable alternative to fossil fuels. By reducing reliance on coal, oil, and natural gas, solar energy can play a critical role in mitigating climate change and improving air quality. This is particularly important in urban areas, where air pollution from fossil fuel combustion is a major public health concern.

The environmental benefits of solar energy extend beyond emissions reductions. Solar installations have a relatively low environmental impact during their production and can be recycled at the end of their lifespan, further enhancing their sustainability. Additionally, solar energy requires minimal water for operation, unlike traditional power plants that rely on water-intensive cooling processes. This makes solar power an attractive option for regions facing water scarcity or drought conditions.

The integration of solar energy into the grid can also enhance energy resilience and reliability. By diversifying the energy mix and reducing dependence on centralized power plants, solar installations can contribute to a more stable and secure energy system. This is particularly important in the face of extreme weather events and natural disasters, which can disrupt traditional energy infrastructure. Distributed solar generation, where electricity is produced close to the point of consumption, can reduce transmission losses and improve grid efficiency.

Despite these benefits, the widespread adoption of solar energy faces several challenges. One of the primary barriers is the intermittent nature of solar power, as electricity generation is dependent on sunlight availability. This variability necessitates the development of energy storage solutions, such as batteries, to store excess energy generated during sunny periods for use during cloudy days or at night. Advances in battery technology and the integration of smart grid systems are essential to overcoming this challenge and ensuring a reliable energy supply.

Policy and regulatory frameworks play a crucial role in promoting the adoption of solar energy. Supportive policies, such as renewable energy targets, carbon pricing, and grid access regulations, can create a favorable environment for solar development. However, these policies must be carefully designed to ensure long-term sustainability and avoid market distortions. Policymakers must also consider the social and economic implications of the energy transition, ensuring that the benefits of solar energy are equitably distributed and that vulnerable communities are not left behind.

Public awareness and education are also vital in overcoming barriers to solar adoption. Many individuals and businesses remain unaware of the benefits and potential of solar energy, leading to misconceptions and hesitancy in embracing solar technology. Educational campaigns and outreach programs can help dispel myths and provide accurate information about the advantages of solar energy, encouraging broader adoption and investment.

The economic and environmental benefits of solar energy make it a cornerstone of the transition to a sustainable energy future. By harnessing the power of the sun, we can reduce our carbon footprint, enhance energy security, and create a more resilient and equitable energy system. As technology continues to advance and costs decline, the potential for solar energy to transform our energy landscape will only grow, offering a path toward a cleaner, more sustainable, and prosperous future.

Overcoming Barriers to Adoption

The adoption of solar energy, while promising, faces several barriers that must be addressed to unlock its full potential. These obstacles range from technological and economic challenges to social and regulatory hurdles. Understanding and overcoming these barriers is essential for individuals, businesses, and policymakers committed to advancing the renewable energy transition.

One of the most significant barriers to solar energy adoption is the initial cost of installation. Although the price of solar panels has decreased dramatically over the years, the upfront investment required for purchasing and installing a solar system can still be prohibitive for many homeowners and small businesses. Financing options, such as solar loans, leases, and power purchase agreements (PPAs), have emerged as viable solutions to mitigate this barrier. These financial instruments allow consumers to spread the cost of solar installations over time, making them more accessible and affordable. Additionally, government incentives, such as tax credits and rebates, can further reduce the financial burden and encourage adoption.

Another challenge is the intermittent nature of solar power, which depends on sunlight availability. This variability can lead to fluctuations in electricity generation, posing a challenge for grid stability and reliability. Energy storage solutions, such as batteries, are critical to addressing this issue by storing excess energy generated during sunny periods for use during cloudy days or at night. Advances in battery technology, including lithium-ion and emerging solid-state batteries, are improving storage capacity and reducing costs, making them increasingly viable for residential and commercial applications. Moreover, integrating solar energy with smart grid technologies can enhance grid management and optimize energy distribution, ensuring a stable and reliable power supply.

Regulatory and policy frameworks play a crucial role in shaping the adoption of solar energy. In some regions, outdated regulations and complex permitting processes can hinder the development of solar projects. Streamlining these processes and implementing supportive policies, such as renewable energy targets and feed-in tariffs, can create a favorable environment for solar adoption. Policymakers must also consider the social and economic implications of the energy transition, ensuring that the benefits of solar energy are equitably distributed and that vulnerable communities are not left behind.

Public awareness and education are vital in overcoming barriers to solar adoption. Many individuals and businesses remain unaware of the benefits and potential of solar energy, leading to misconceptions and hesitancy in embracing solar technology.

Educational campaigns and outreach programs can help dispel myths and provide accurate information about the advantages of solar energy, encouraging broader adoption and investment. By highlighting success stories and showcasing the tangible benefits of solar installations, these initiatives can inspire confidence and motivate action.

The integration of solar energy into urban environments presents unique challenges and opportunities. Cities, with their dense populations and high energy demands, offer significant potential for solar energy generation. However, the limited availability of open space and the presence of tall buildings can complicate the installation of solar panels. Innovative solutions, such as building-integrated photovoltaics (BIPV) and community solar projects, can overcome these obstacles by incorporating solar technology into the fabric of the city. BIPV systems, which integrate solar cells directly into building materials, offer aesthetic appeal and space efficiency, while community solar projects allow residents and businesses to collectively invest in and benefit from solar energy.

The environmental impact of solar installations is another consideration that must be addressed. While solar energy is a clean and sustainable alternative to fossil fuels, the production and disposal of solar panels can have environmental consequences. To mitigate these impacts, the solar industry is increasingly focusing on sustainable manufacturing practices and recycling initiatives. By developing closed-loop systems that recover valuable materials from end-of-life panels, the industry can reduce waste and minimize its environmental footprint.

The social acceptance of solar energy is also a critical factor in its adoption. In some communities, concerns about the visual impact of solar installations or potential land use conflicts can lead to resistance. Engaging with local communities and stakeholders early in the planning process can help address these concerns and build support for solar projects. By involving community members in decision-making and demonstrating the benefits of solar energy, developers can foster a sense of ownership and collaboration.

The role of innovation in overcoming barriers to solar adoption cannot be overstated. Technological advancements, such as the development of high-efficiency solar cells and novel materials, are driving down costs and improving performance. Research into new solar technologies, such as perovskite solar cells and organic photovoltaics, holds the promise of further breakthroughs that could revolutionize the industry. By investing in research and development, governments and private companies can accelerate the pace of innovation and unlock new opportunities for solar energy.

Collaboration and partnerships are essential to overcoming barriers and driving the adoption of solar energy. By working together, governments, businesses, and communities can share knowledge, resources, and best practices, creating synergies that amplify the impact of individual efforts. International cooperation is also crucial, as the global nature of the energy transition requires coordinated action and the sharing of experiences across borders.

The transition to solar energy is a complex and multifaceted process that requires a holistic approach to address the various barriers to adoption. By leveraging innovative solutions, supportive policies, and collaborative efforts, we can overcome these challenges and unlock the full potential of solar energy. As technology continues to advance and costs decline, the path toward a sustainable and resilient energy future becomes increasingly attainable. The journey may be challenging, but the rewards of a cleaner, more sustainable world are well worth the effort.

Chapter 2: Wind Energy: Capturing Nature's Breeze

The Mechanics of Wind Turbines

Wind turbines, towering structures that dot landscapes and coastlines, are marvels of engineering designed to harness the kinetic energy of the wind and convert it into electricity. Understanding the mechanics of these turbines is essential for appreciating their role in the renewable energy landscape and for those considering their implementation.

At the heart of a wind turbine is the rotor, which consists of blades attached to a central hub. These blades are aerodynamically designed to capture the wind's energy. As the wind blows, it creates lift on the blades, similar to the way an airplane wing generates lift. This lift causes the rotor to spin, converting the linear motion of the wind into rotational energy. The number of blades on a turbine can vary, but most modern turbines use three blades, which offer a balance between efficiency and mechanical stability.

The rotor is connected to a shaft, which transfers the rotational energy to a generator housed within the nacelle, the large casing atop the turbine tower. The generator is a critical component, as it is responsible for converting the mechanical energy from the rotor into electrical energy. Most wind turbines use an induction generator, which operates on the principle of electromagnetic induction. As the rotor turns the shaft, it spins a set of magnets within the generator, inducing an electrical current in the surrounding coils of wire.

To optimize energy capture, wind turbines are equipped with a yaw system, which allows the nacelle to rotate and face the wind. This system uses sensors to detect the wind's direction and motors to adjust the nacelle's position, ensuring that the rotor is always aligned with the wind for maximum efficiency. Additionally, the pitch of the blades can be adjusted to control the rotor's speed and optimize performance in varying wind conditions. This is achieved through a pitch control system, which uses hydraulic or electric actuators to change the angle of the blades relative to the wind.

The tower of a wind turbine serves as the structural backbone, supporting the nacelle and rotor at a height where wind speeds are typically higher and more consistent. Towers are usually constructed from steel or concrete and can reach heights of over 100 meters. The height of the tower is a critical factor in the turbine's performance, as wind speed increases with altitude. Taller towers can access stronger winds, resulting in greater energy output.

Wind turbines are equipped with a range of safety and control systems to ensure reliable operation and protect the equipment from damage. One such system is the brake, which can be used to stop the rotor in high wind conditions or during maintenance. The brake can be mechanical, using friction to halt the rotor, or aerodynamic, using the pitch control system to feather the blades and reduce lift. Additionally, turbines are equipped with anemometers and wind vanes to monitor wind speed and direction, providing data for the control systems to adjust the turbine's operation.

The efficiency of a wind turbine is influenced by several factors, including the design of the blades, the height of the tower, and the location of the installation. The Betz limit, a theoretical maximum efficiency of 59.3%, represents the upper bound of energy that can be extracted from the wind by a turbine. While no turbine can achieve this limit, modern designs can reach efficiencies of 35% to 45%, making them a viable source of renewable energy.

The siting of wind turbines is a critical consideration, as the availability and consistency of wind resources directly impact their performance. Ideal locations for wind farms are areas with high average wind speeds and minimal turbulence, such as open plains, coastal regions, and offshore sites. The layout of a wind farm is also important, as turbines must be spaced to minimize wake effects, where the turbulence created by one turbine can reduce the efficiency of others downwind.

Offshore wind turbines present unique challenges and opportunities. While they can access stronger and more consistent winds than onshore installations, they require specialized foundations and construction techniques to withstand the harsh marine environment. Offshore turbines are typically larger than their onshore counterparts, with capacities exceeding 10 megawatts, allowing them to generate significant amounts of electricity. The development of floating wind turbines, which can be anchored in deeper waters, is expanding the potential for offshore wind energy.

The integration of wind energy into the electrical grid requires careful management to balance supply and demand. Wind power is inherently variable, as it depends on weather conditions, necessitating the use of complementary energy sources and storage solutions to ensure a stable and reliable electricity supply. Advances in grid management technologies, such as smart grids and demand response systems, are helping to address these challenges and facilitate the integration of wind energy.

The environmental impact of wind turbines is generally positive, as they produce electricity without emitting greenhouse gases or air pollutants. However, concerns have been raised about their effects on wildlife, particularly birds and bats, which can collide with the blades. To mitigate these impacts, developers are exploring strategies such as siting turbines away from migration routes, using radar systems to detect and deter approaching wildlife, and designing blades with features that reduce collision risk.

Public perception and acceptance of wind turbines can influence their deployment. While many people support wind energy for its environmental benefits, others may have concerns about visual impact, noise, and land use. Engaging with local communities and stakeholders is essential to address these concerns and build support for wind projects. By involving community members in the planning process and demonstrating the benefits of wind energy, developers can foster a sense of ownership and collaboration.

The mechanics of wind turbines represent a fascinating intersection of engineering, physics, and environmental science. By harnessing the power of the wind, these machines offer a sustainable and scalable solution to our energy needs. As technology continues to advance, the potential for wind energy to contribute to a cleaner and more resilient energy future will only grow, providing a path toward a more sustainable world.

Offshore vs. Onshore Wind Farms

The debate between offshore and onshore wind farms is a pivotal one in the realm of renewable energy, each offering distinct advantages and challenges. Understanding these differences is crucial for stakeholders, from policymakers to investors, as they make informed decisions about the future of wind energy.

Onshore wind farms, located on land, have been the traditional choice for wind energy projects. They are generally easier and less expensive to install compared to their offshore counterparts. The infrastructure required for onshore wind farms, such as roads and power lines, is typically more accessible and less costly to develop. This accessibility translates to lower initial capital investment, making onshore projects attractive to developers and investors.

The maintenance and operation of onshore wind farms are also more straightforward. Technicians can access turbines by road, reducing the logistical complexities and costs associated with repairs and regular maintenance. This ease of access contributes to lower operational expenses over the lifespan of the wind farm, enhancing its economic viability.

However, onshore wind farms face limitations in terms of location and public acceptance. Suitable sites for onshore wind farms are often in remote or rural areas, where wind speeds are higher and more consistent. Yet, these locations can be far from urban centers where electricity demand is greatest, necessitating the construction of extensive transmission lines. Additionally, the visual and noise impact of onshore turbines can lead to opposition from local communities, complicating the permitting process and potentially delaying projects.

Offshore wind farms, situated in bodies of water, offer a compelling alternative with their own set of benefits. One of the most significant advantages of offshore wind farms is the availability of stronger and more consistent winds. The open sea provides an unobstructed path for wind currents, resulting in higher energy yields compared to onshore sites. This increased efficiency can offset the higher costs associated with offshore installations, making them economically competitive in the long run.

The location of offshore wind farms also presents opportunities for proximity to densely populated coastal regions. By situating wind farms near urban centers, the need for long-distance transmission lines is reduced, minimizing energy losses and infrastructure costs. This proximity can enhance the reliability and stability of the electricity supply, particularly in regions with high energy demands.

Despite these advantages, offshore wind farms face significant challenges, primarily related to installation and maintenance. The harsh marine environment requires specialized equipment and techniques for constructing and anchoring turbines. Foundations must be robust enough to withstand strong currents, waves, and corrosive saltwater, adding to the complexity and cost of offshore projects. Furthermore, accessing offshore turbines for maintenance and repairs can be logistically challenging and weather-dependent, leading to higher operational costs.

The environmental impact of both onshore and offshore wind farms is a critical consideration in their development. Onshore wind farms can affect local ecosystems, particularly if they are located in sensitive habitats or migration corridors. Careful site selection and environmental assessments are essential to minimize these impacts and ensure the sustainability of onshore projects.

Offshore wind farms, while generally having a lower visual impact, can affect marine life and ecosystems. The construction and operation of offshore turbines can disrupt marine habitats and pose risks to birds and marine mammals. Mitigation measures, such as careful site selection, monitoring, and the use of technology to deter wildlife, are necessary to address these concerns and protect marine biodiversity.

The economic implications of onshore and offshore wind farms extend beyond their immediate costs and benefits. Both types of projects contribute to job creation and economic development, supporting local communities and industries. Onshore wind farms often provide employment opportunities in rural areas, where economic development may be limited. Offshore wind farms, with their complex infrastructure and maintenance needs, can drive innovation and growth in the maritime and engineering sectors.

Policy and regulatory frameworks play a crucial role in shaping the development of onshore and offshore wind farms. Supportive policies, such as subsidies, tax incentives, and renewable energy targets, can create a favorable environment for investment and development. Policymakers must balance the need for rapid deployment of wind energy with the protection of natural and cultural resources, ensuring that projects are sustainable and socially acceptable.

Public perception and acceptance are vital to the success of wind energy projects. Engaging with local communities and stakeholders early in the planning process can help address concerns and build support for both onshore and offshore projects. Transparent communication, community involvement, and the demonstration of tangible benefits, such as job creation and local investment, can foster a sense of ownership and collaboration.

The choice between onshore and offshore wind farms is not a simple one, as each offers unique advantages and challenges. The decision often depends on a range of factors, including geographic location, wind resources, economic considerations, and environmental impacts. By carefully evaluating these factors and leveraging the strengths of both onshore and offshore wind energy, stakeholders can develop a balanced and sustainable approach to harnessing the power of the wind.

As technology continues to advance, the potential for both onshore and offshore wind farms to contribute to a cleaner and more resilient energy future will only grow. Innovations in turbine design, materials, and installation techniques are driving down costs and improving performance, making wind energy an increasingly attractive option for meeting global energy needs. The journey toward a sustainable energy future is complex, but the rewards of harnessing the power of the wind are immense, offering a path toward a cleaner, more sustainable world.

Technological Advancements in Wind Power

The evolution of wind power technology has been nothing short of revolutionary, transforming it from a niche energy source into a cornerstone of the global renewable energy landscape. Technological advancements have played a pivotal role in enhancing the efficiency, reliability, and cost-effectiveness of wind energy, making it a viable alternative to fossil fuels. These innovations span various aspects of wind power, from turbine design and materials to energy storage and grid integration.

One of the most significant advancements in wind power technology is the development of larger and more efficient wind turbines. Modern turbines have grown in size, with rotor diameters exceeding 150 meters and tower heights reaching over 100 meters. This increase in size allows turbines to capture more wind energy, particularly at higher altitudes where wind speeds are greater and more consistent. The result is a substantial boost in energy output, making wind farms more productive and economically viable.

The design of wind turbine blades has also seen remarkable progress. Engineers have employed advanced materials, such as carbon fiber composites, to create lighter and more durable blades. These materials not only enhance the structural integrity of the blades but also improve their aerodynamic performance, allowing them to capture more energy from the wind. Additionally, innovations in blade design, such as curved and segmented blades, have further optimized energy capture and reduced noise levels, addressing some of the common concerns associated with wind turbines.

The integration of smart technologies into wind turbines has revolutionized their operation and maintenance. Sensors and data analytics are now used to monitor turbine performance in real-time, enabling predictive maintenance and reducing downtime. By analyzing data on wind conditions, vibration, and temperature, operators can identify potential issues before they lead to costly failures. This proactive approach not only extends the lifespan of turbines but also enhances their reliability and efficiency.

Energy storage solutions have emerged as a critical component of wind power systems, addressing the inherent variability of wind energy. Advances in battery technology, such as lithium-ion and flow batteries, have improved storage capacity and reduced costs, making them increasingly viable for large-scale applications. These storage systems allow excess energy generated during periods of high wind to be stored and used when wind speeds are low, ensuring a stable and reliable electricity supply. Furthermore, the development of hybrid systems that combine wind power with other renewable sources, such as solar, has enhanced the overall resilience and flexibility of the energy grid.

Grid integration technologies have also advanced significantly, facilitating the seamless incorporation of wind energy into existing power systems. Smart grid technologies, which use digital communication and automation, enable more efficient management of electricity supply and demand. By dynamically adjusting the flow of electricity based on real-time data, smart grids can accommodate the variability of wind energy and maintain grid stability. Additionally, advances in power electronics, such as inverters and transformers, have improved the efficiency of energy conversion and transmission, reducing losses and enhancing the overall performance of wind power systems.

Offshore wind technology has seen notable advancements, expanding the potential for wind energy generation. The development of floating wind turbines, which can be anchored in deeper waters, has opened up new opportunities for offshore wind farms. These floating platforms are designed to withstand harsh marine conditions, allowing them to access stronger and more consistent winds far from the coast. The increased energy yield from offshore installations can offset the higher costs associated with their construction and maintenance, making them an attractive option for expanding wind energy capacity.

The use of artificial intelligence and machine learning in wind power is another area of rapid development. These technologies are being used to optimize turbine performance, predict maintenance needs, and improve energy forecasting. By analyzing vast amounts of data from wind farms, AI algorithms can identify patterns and trends that inform decision-making and enhance operational efficiency. This data-driven approach not only maximizes energy output but also reduces costs and minimizes environmental impact.

The environmental impact of wind power technology has been a focus of ongoing research and innovation. Efforts to minimize the ecological footprint of wind farms include the development of wildlife-friendly turbine designs and the implementation of monitoring systems to protect birds and bats. Additionally, advances in recycling and end-of-life management for turbine components are addressing concerns about waste and resource use, ensuring that wind power remains a sustainable and environmentally responsible energy source.

Public perception and acceptance of wind power technology are crucial for its continued growth and development. Engaging with local communities and stakeholders is essential to address concerns and build support for wind projects. Transparent communication, community involvement, and the demonstration of tangible benefits, such as job creation and local investment, can foster a sense of ownership and collaboration.

The future of wind power technology is bright, with ongoing research and development promising even greater advancements. Innovations in materials science, aerodynamics, and digital technologies are set to further enhance the efficiency and cost-effectiveness of wind energy. As the global demand for clean and sustainable energy continues to rise, wind power will play an increasingly important role in meeting this demand and driving the transition to a low-carbon economy.

The journey of wind power technology is a testament to human ingenuity and the relentless pursuit of progress. By harnessing the power of the wind, we are not only addressing the urgent challenges of climate change and energy security but also paving the way for a more sustainable and prosperous future. The potential of wind energy is vast, and with continued innovation and collaboration, we can unlock its full potential and create a cleaner, more resilient world for generations to come.

Environmental and Economic Impacts

The dual forces of environmental and economic impacts shape the narrative of renewable energy, particularly in the context of wind power. As the world grapples with the pressing need to transition to sustainable energy sources, understanding these impacts is crucial for informed decision-making and strategic planning.

Wind power stands out as a clean and sustainable energy source, offering significant environmental benefits. Unlike fossil fuels, wind energy generates electricity without emitting greenhouse gases or air pollutants, making it a vital tool in combating climate change. By displacing carbon-intensive energy sources, wind power contributes to reducing the global carbon footprint, helping nations meet their emissions reduction targets and mitigate the adverse effects of climate change.

The environmental benefits of wind power extend beyond carbon emissions. Wind turbines require no water for operation, unlike conventional power plants that consume vast quantities of water for cooling. This characteristic makes wind energy particularly valuable in arid regions where water scarcity is a pressing concern. Additionally, wind farms occupy relatively small land areas compared to other energy sources, allowing for dual land use, such as agriculture or grazing, beneath the turbines.

However, the deployment of wind power is not without its environmental challenges. The construction and operation of wind farms can impact local ecosystems and wildlife. Birds and bats, in particular, are at risk of collision with turbine blades, leading to concerns about biodiversity conservation. To address these issues, developers are implementing mitigation measures, such as careful site selection, turbine design modifications, and the use of technology to detect and deter wildlife. Ongoing research and monitoring are essential to minimize these impacts and ensure the sustainability of wind energy projects.

The economic impacts of wind power are multifaceted, influencing job creation, energy prices, and local economies. The wind energy sector has emerged as a significant source of employment, supporting jobs in manufacturing, construction, operation, and maintenance. As the industry continues to grow, it is expected to create even more job opportunities, contributing to economic development and workforce diversification.

Wind power also has the potential to stabilize and reduce energy prices. Once a wind farm is operational, the cost of generating electricity is relatively low, as wind is a free and abundant resource. This characteristic makes wind energy less susceptible to price volatility compared to fossil fuels, which are subject to market fluctuations and geopolitical tensions. By diversifying the energy mix and reducing reliance on imported fuels, wind power can enhance energy security and provide a buffer against price shocks.

The economic benefits of wind power extend to local communities, particularly in rural areas where wind farms are often located. Landowners can receive lease payments for hosting turbines on their property, providing a steady source of income. Additionally, wind farms contribute to local tax revenues, which can be used to fund public services and infrastructure improvements. These economic benefits can foster community support for wind projects and promote regional development.

Despite these advantages, the economic viability of wind power is influenced by several factors, including government policies, market conditions, and technological advancements. Supportive policies, such as subsidies, tax incentives, and renewable energy targets, play a crucial role in creating a favorable environment for wind energy investment. Policymakers must balance the need for financial support with the goal of fostering a competitive and sustainable energy market.

The cost of wind power has decreased significantly over the past decade, driven by technological advancements and economies of scale. Innovations in turbine design, materials, and manufacturing processes have improved efficiency and reduced costs, making wind energy increasingly competitive with conventional power sources. As technology continues to advance, the cost of wind power is expected to decline further, enhancing its economic attractiveness.

The integration of wind power into the energy grid presents both opportunities and challenges. The variability of wind energy requires complementary energy sources and storage solutions to ensure a stable and reliable electricity supply. Advances in grid management technologies, such as smart grids and demand response systems, are helping to address these challenges and facilitate the integration of wind energy. By enhancing grid flexibility and resilience, these technologies can maximize the economic and environmental benefits of wind power.

Public perception and acceptance of wind power are critical to its continued growth and development. While many people support wind energy for its environmental and economic benefits, others may have concerns about visual impact, noise, and land use. Engaging with local communities and stakeholders is essential to address these concerns and build support for wind projects. Transparent communication, community involvement, and the demonstration of tangible benefits can foster a sense of ownership and collaboration.

The environmental and economic impacts of wind power are interconnected, shaping the future of renewable energy. By harnessing the power of the wind, we can address the urgent challenges of climate change and energy security while promoting economic development and sustainability. The journey toward a cleaner and more resilient energy future is complex, but the rewards of wind power are immense, offering a path toward a more sustainable and prosperous world. As technology continues to advance and costs decline, the potential for wind energy to contribute to a cleaner and more resilient energy future will only grow, providing a path toward a more sustainable world.

Policy and Regulatory Frameworks

Navigating the complex landscape of policy and regulatory frameworks is essential for the successful deployment and expansion of wind energy projects. These frameworks shape the development, financing, and operation of wind farms, influencing everything from site selection to grid integration. Understanding the intricacies of these policies is crucial for stakeholders, including developers, investors, and policymakers, as they work to harness the potential of wind energy.

At the core of wind energy policy are renewable energy targets, which set specific goals for the proportion of energy to be generated from renewable sources. These targets provide a clear signal to the market, encouraging investment in wind energy and other renewables. By establishing ambitious yet achievable targets, governments can drive the transition to a low-carbon energy system and stimulate economic growth in the renewable energy sector.

Subsidies and financial incentives play a pivotal role in supporting the development of wind energy projects. These incentives can take various forms, including feed-in tariffs, tax credits, and grants, each designed to reduce the financial risk and enhance the economic viability of wind projects. Feed-in tariffs guarantee a fixed price for electricity generated from wind power, providing a stable revenue stream for developers. Tax credits, on the other hand, reduce the tax liability of wind energy companies, improving their financial performance and attractiveness to investors.

Permitting and licensing processes are critical components of the regulatory framework, governing the approval and construction of wind farms. These processes ensure that projects comply with environmental, social, and technical standards, balancing the need for renewable energy development with the protection of natural and cultural resources. Streamlining permitting procedures can reduce delays and costs, facilitating the timely deployment of wind energy projects. However, it is essential to maintain rigorous standards to safeguard environmental and community interests.

Grid integration policies are vital for the successful incorporation of wind energy into the existing power system. These policies address the technical and operational challenges associated with the variability of wind power, ensuring a stable and reliable electricity supply. Grid codes and standards define the technical requirements for connecting wind farms to the grid, covering aspects such as voltage, frequency, and reactive power. By establishing clear and consistent grid integration policies, regulators can enhance the efficiency and reliability of the power system, maximizing the benefits of wind energy.

Land use and zoning regulations are another critical aspect of the policy framework, influencing the siting and development of wind farms. These regulations determine where wind projects can be located, taking into account factors such as land ownership, environmental sensitivity, and community preferences. By carefully balancing the needs of wind energy development with land use considerations, policymakers can minimize conflicts and promote sustainable land management practices.

Environmental impact assessments (EIAs) are a key regulatory requirement for wind energy projects, evaluating the potential effects of a proposed development on the environment. EIAs provide a comprehensive analysis of the project's impact on wildlife, habitats, and ecosystems, informing decision-making and ensuring that projects are environmentally responsible. By identifying potential risks and mitigation measures, EIAs help to minimize the environmental footprint of wind farms and promote sustainable development.

Public engagement and stakeholder consultation are essential components of the policy and regulatory framework, fostering transparency and collaboration in the development of wind energy projects. By involving local communities, landowners, and other stakeholders in the planning process, developers can address concerns, build trust, and secure support for their projects. Effective communication and consultation can lead to mutually beneficial outcomes, enhancing the social acceptance and success of wind energy initiatives.

International cooperation and harmonization of policies are increasingly important in the global wind energy landscape. Cross-border collaboration can facilitate the sharing of best practices, technologies, and resources, accelerating the deployment of wind energy worldwide. International agreements and frameworks, such as the Paris Agreement, provide a platform for countries to coordinate their efforts and align their policies with global climate and energy goals.

The role of government and regulatory bodies in shaping the policy framework for wind energy cannot be overstated. These entities are responsible for setting the strategic direction, establishing regulatory standards, and ensuring compliance with legal and environmental requirements. By providing clear and consistent guidance, governments can create a stable and predictable environment for wind energy investment and development.

The dynamic nature of the wind energy sector requires continuous adaptation and evolution of policy and regulatory frameworks. As technology advances and market conditions change, policymakers must remain agile and responsive, updating policies to reflect new realities and opportunities. By fostering innovation and flexibility, regulatory frameworks can support the growth and maturation of the wind energy industry, driving the transition to a sustainable energy future.

The interplay between policy, regulation, and wind energy development is complex and multifaceted, requiring a holistic and integrated approach. By understanding and navigating the policy and regulatory landscape, stakeholders can unlock the full potential of wind energy, contributing to a cleaner, more resilient, and sustainable energy system. The journey toward a low-carbon future is challenging, but with the right policies and frameworks in place, wind energy can play a central role in achieving global climate and energy goals.

Chapter 3: Hydropower: The Force of Water

Understanding Hydroelectric Systems

Hydroelectric systems have long been a cornerstone of renewable energy, harnessing the power of flowing water to generate electricity. These systems, which convert the kinetic energy of water into mechanical energy and then into electrical energy, have been utilized for over a century and continue to play a vital role in the global energy mix. Understanding the intricacies of hydroelectric systems is essential for appreciating their benefits, challenges, and potential for future development.

At the heart of a hydroelectric system is the dam, a structure built across a river to control the flow of water. By creating a reservoir, the dam stores water and releases it in a controlled manner to generate electricity. The height of the dam and the volume of water in the reservoir determine the potential energy available for conversion into electricity. This potential energy is transformed into kinetic energy as water flows through the dam's turbines, which spin to drive generators and produce electricity.

There are several types of hydroelectric systems, each with its unique characteristics and applications. The most common type is the impoundment facility, which uses a dam to store water in a reservoir. This stored water can be released as needed to generate electricity, providing a reliable and flexible source of power. Impoundment facilities are often large-scale projects, capable of generating significant amounts of electricity and providing flood control, irrigation, and recreational opportunities.

Another type of hydroelectric system is the run-of-river facility, which generates electricity without the need for a large reservoir. Instead, these systems divert a portion of a river's flow through a channel or penstock to drive turbines. Run-of-river facilities have a smaller environmental footprint compared to impoundment facilities, as they do not require extensive flooding of land. However, their electricity generation capacity is more variable, depending on the river's flow and seasonal changes.

Pumped storage is a specialized type of hydroelectric system that acts as a large-scale energy storage solution. During periods of low electricity demand, excess energy is used to pump water from a lower reservoir to an upper reservoir. When demand increases, the stored water is released back to the lower reservoir, generating electricity in the process. Pumped storage systems provide grid stability and flexibility, balancing supply and demand and supporting the integration of other renewable energy sources.

The environmental impacts of hydroelectric systems are a critical consideration in their development and operation. While hydroelectric power is a clean and renewable energy source, the construction of dams and reservoirs can have significant ecological consequences. Flooding large areas of land can disrupt ecosystems, displace wildlife, and alter natural water flow patterns. These changes can affect fish populations, water quality, and sediment transport, with potential downstream impacts on agriculture and communities.

To mitigate these environmental impacts, developers and operators implement various strategies and technologies. Fish ladders and bypass systems are designed to help fish navigate around dams, maintaining natural migration patterns and supporting biodiversity. Environmental flow releases, which mimic natural river flow patterns, help maintain downstream ecosystems and water quality. Additionally, careful site selection and environmental assessments are essential to minimize the ecological footprint of hydroelectric projects.

The economic benefits of hydroelectric systems are substantial, contributing to energy security, job creation, and regional development. Hydroelectric power provides a stable and reliable source of electricity, reducing dependence on fossil fuels and enhancing energy security. The construction and operation of hydroelectric facilities create jobs in engineering, construction, and maintenance, supporting local economies and workforce development.

Hydroelectric systems also offer long-term economic advantages, as they have low operating and maintenance costs once constructed. The longevity of hydroelectric facilities, which can operate for several decades, provides a consistent return on investment and contributes to the overall affordability of electricity. Furthermore, the multipurpose nature of many hydroelectric projects, which provide benefits such as flood control, irrigation, and recreation, enhances their economic value and social acceptance.

The integration of hydroelectric power into the energy grid presents both opportunities and challenges. Hydroelectric systems provide grid stability and flexibility, as they can quickly adjust their output to match changes in electricity demand. This capability makes them an ideal complement to other renewable energy sources, such as wind and solar, which are more variable in nature. However, the variability of water flow, influenced by factors such as climate change and competing water uses, can affect the reliability of hydroelectric power.

Advancements in technology and innovation are driving the evolution of hydroelectric systems, enhancing their efficiency and sustainability. Improvements in turbine design and materials have increased the efficiency of energy conversion, maximizing electricity generation from available water resources. Digital technologies, such as sensors and data analytics, are being used to optimize the operation and maintenance of hydroelectric facilities, reducing costs and improving performance.

The future of hydroelectric systems is shaped by a range of factors, including technological advancements, environmental considerations, and policy frameworks. As the global demand for clean and sustainable energy continues to grow, hydroelectric power will play an increasingly important role in meeting this demand. By balancing the benefits and challenges of hydroelectric systems, stakeholders can develop strategies that maximize their potential and contribute to a sustainable energy future.

The journey of hydroelectric systems is a testament to the power of innovation and the potential of renewable energy. By harnessing the energy of flowing water, we can address the urgent challenges of climate change and energy security while promoting economic development and environmental sustainability. The potential of hydroelectric power is vast, and with continued innovation and collaboration, we can unlock its full potential and create a cleaner, more resilient world for generations to come.

Large-Scale Dams vs. Small Hydro Projects

The debate between large-scale dams and small hydro projects is a nuanced one, reflecting the diverse needs and priorities of energy systems worldwide. Both approaches to hydroelectric power generation offer unique advantages and challenges, and understanding these differences is crucial for making informed decisions about energy development and sustainability.

Large-scale dams, often referred to as mega-dams, are monumental engineering feats that have the capacity to generate vast amounts of electricity. These structures, such as the Hoover Dam in the United States or the Three Gorges Dam in China, are capable of producing gigawatts of power, making them significant contributors to national energy grids. The sheer scale of these projects allows them to provide a stable and reliable source of electricity, which can support industrial activities, urban centers, and regional development.

One of the primary advantages of large-scale dams is their ability to store substantial volumes of water, creating reservoirs that can be used for multiple purposes. These reservoirs not only generate electricity but also provide water for irrigation, flood control, and recreational activities. The multipurpose nature of large-scale dams can enhance their economic value and social acceptance, as they offer a range of benefits beyond energy generation.

However, the construction and operation of large-scale dams come with significant environmental and social challenges. The creation of reservoirs often requires the flooding of extensive land areas, which can disrupt ecosystems, displace communities, and alter natural water flow patterns. The environmental impact of large-scale dams can be profound, affecting biodiversity, water quality, and sediment transport. Additionally, the social implications of displacing communities and altering traditional land uses can lead to conflicts and resistance.

In contrast, small hydro projects, also known as small-scale or micro-hydro systems, offer a more localized and environmentally friendly approach to hydroelectric power generation. These projects typically generate less than 10 megawatts of electricity and are often designed to serve specific communities or regions. Small hydro projects can be implemented with minimal environmental disruption, as they do not require large reservoirs or extensive land flooding.

The localized nature of small hydro projects allows them to be tailored to the specific needs and conditions of a community or region. They can provide a reliable source of electricity for remote or off-grid areas, supporting local development and improving quality of life. Small hydro projects can also be integrated with other renewable energy sources, such as solar or wind, to create hybrid systems that enhance energy resilience and sustainability.

Despite their advantages, small hydro projects face challenges related to scale and economic viability. The limited capacity of these projects means they may not be suitable for meeting the energy demands of large urban centers or industrial activities.

Additionally, the cost per unit of electricity generated by small hydro projects can be higher than that of large-scale dams, due to economies of scale and the need for specialized equipment and infrastructure.

The choice between large-scale dams and small hydro projects is influenced by a range of factors, including geographic conditions, energy needs, environmental considerations, and policy frameworks. In regions with abundant water resources and high energy demands, large-scale dams may be the preferred option, offering a significant and stable source of electricity. However, in areas with sensitive ecosystems or limited water availability, small hydro projects may provide a more sustainable and adaptable solution.

Policy and regulatory frameworks play a crucial role in shaping the development of both large-scale dams and small hydro projects. Supportive policies, such as subsidies, tax incentives, and streamlined permitting processes, can encourage investment and innovation in hydroelectric power. Policymakers must balance the need for renewable energy development with the protection of environmental and social values, ensuring that projects are sustainable and equitable.

Technological advancements are driving the evolution of both large-scale dams and small hydro projects, enhancing their efficiency and sustainability. Innovations in turbine design, materials, and digital technologies are improving the performance and reducing the environmental impact of hydroelectric systems. These advancements are enabling the development of more efficient and adaptable hydroelectric solutions, capable of meeting the diverse needs of modern energy systems.

The integration of hydroelectric power into the broader energy landscape requires careful consideration of its interactions with other energy sources and grid infrastructure. Large-scale dams can provide grid stability and flexibility, complementing variable renewable sources such as wind and solar. Small hydro projects can support decentralized energy systems, enhancing energy access and resilience in remote or underserved areas.

The future of hydroelectric power lies in the ability to balance the benefits and challenges of large-scale dams and small hydro projects. By understanding the unique characteristics and potential of each approach, stakeholders can develop strategies that maximize the contributions of hydroelectric power to a sustainable energy future. Collaboration, innovation, and adaptive management are key to unlocking the full potential of hydroelectric systems and creating a cleaner, more resilient world.

The journey of hydroelectric power is a testament to the power of human ingenuity and the potential of renewable energy. By harnessing the energy of flowing water, we can address the urgent challenges of climate change and energy security while promoting economic development and environmental sustainability. The potential of hydroelectric power is vast, and with continued innovation and collaboration, we can unlock its full potential and create a cleaner, more resilient world for generations to come.

Environmental Considerations and Mitigation

The interplay between energy development and environmental stewardship is a delicate balance, particularly in the realm of renewable energy. As the world increasingly turns to sustainable sources like wind, solar, and hydroelectric power, understanding the environmental considerations and implementing effective mitigation strategies becomes paramount. These efforts ensure that the transition to cleaner energy does not come at the expense of ecological integrity.

Renewable energy projects, while inherently more environmentally friendly than fossil fuel counterparts, are not without their ecological footprints. The construction and operation of these projects can impact local ecosystems, wildlife, and natural resources. For instance, wind farms, though emission-free, pose risks to avian and bat populations due to turbine collisions. Similarly, solar farms require large tracts of land, which can disrupt habitats and alter land use patterns. Hydroelectric projects, particularly large-scale dams, can significantly alter river ecosystems, affecting fish migration and water quality.

Addressing these environmental impacts requires a comprehensive approach that begins with careful site selection and planning. Identifying locations that minimize ecological disruption is a critical first step. This involves conducting thorough environmental impact assessments (EIAs) to evaluate potential effects on wildlife, habitats, and natural resources. EIAs provide valuable insights into the ecological context of a project, informing decision-making and guiding the development of mitigation strategies.

Mitigation measures are essential for minimizing the environmental impacts of renewable energy projects. For wind farms, technological innovations such as radar systems and ultrasonic deterrents can help reduce bird and bat collisions.

Adjusting turbine operation during peak migration periods is another effective strategy. In the case of solar farms, integrating vegetation management practices and designing wildlife corridors can help preserve local biodiversity. For hydroelectric projects, fish ladders and bypass systems facilitate the movement of aquatic species, maintaining natural migration patterns and supporting ecosystem health.

Community engagement and stakeholder involvement are crucial components of environmental mitigation. By involving local communities, landowners, and environmental organizations in the planning process, developers can address concerns, build trust, and foster collaboration. Transparent communication and participatory decision-making can lead to mutually beneficial outcomes, enhancing the social acceptance and success of renewable energy projects.

Adaptive management is a key principle in the ongoing stewardship of renewable energy projects. This approach involves monitoring environmental impacts over time and adjusting mitigation measures as needed. By continuously evaluating the effectiveness of mitigation strategies, developers can respond to changing conditions and emerging challenges, ensuring that projects remain environmentally sustainable.

Policy and regulatory frameworks play a vital role in shaping environmental considerations and mitigation efforts. Supportive policies, such as environmental standards, permitting requirements, and incentives for sustainable practices, create a conducive environment for responsible energy development. Policymakers must balance the need for renewable energy expansion with the protection of environmental and social values, ensuring that projects are both economically viable and ecologically sound.

Technological advancements are driving innovation in environmental mitigation, enhancing the ability to minimize impacts and improve sustainability. For example, advancements in turbine design and materials are reducing the noise and visual impact of wind farms. In solar energy, the development of bifacial panels and floating solar arrays is increasing efficiency and reducing land use requirements. In hydroelectric power, improvements in turbine technology are enhancing fish passage and reducing water quality impacts.

The integration of renewable energy into the broader energy landscape requires careful consideration of its interactions with natural systems and resources. By understanding the environmental context and implementing effective mitigation strategies, stakeholders can maximize the benefits of renewable energy while minimizing its ecological footprint. Collaboration, innovation, and adaptive management are key to achieving a sustainable energy future that respects and preserves the natural world.

The journey toward a cleaner and more sustainable energy system is complex, but the rewards are immense. By harnessing the power of renewable energy, we can address the urgent challenges of climate change and energy security while promoting economic development and environmental sustainability. The potential of renewable energy is vast, and with continued innovation and collaboration, we can unlock its full potential and create a cleaner, more resilient world for generations to come.

The Role of Hydropower in Energy Grids

Hydropower has long been a cornerstone of the global energy landscape, providing a reliable and renewable source of electricity that plays a crucial role in modern energy grids. As the world seeks to transition to cleaner energy sources, understanding the role of hydropower within energy grids is essential for optimizing its benefits and addressing the challenges it presents.

Hydropower's contribution to energy grids is multifaceted, offering both base-load and peak-load power generation capabilities. Base-load power refers to the continuous supply of electricity needed to meet the minimum demand on the grid. Hydropower plants, particularly large-scale dams, are well-suited for this role due to their ability to generate a steady and reliable flow of electricity. This stability is a significant advantage, as it ensures a consistent supply of power to homes, businesses, and industries.

In addition to base-load power, hydropower is also adept at providing peak-load power, which is the additional electricity required during periods of high demand. The flexibility of hydropower plants allows them to quickly ramp up or down their output, making them ideal for meeting sudden spikes in electricity consumption. This capability is particularly valuable in energy grids that incorporate variable renewable energy sources, such as wind and solar, which can fluctuate based on weather conditions.

The integration of hydropower into energy grids enhances grid stability and resilience. By providing a reliable source of electricity, hydropower helps to balance supply and demand, reducing the risk of blackouts and ensuring a stable power supply. This stability is further supported by the ability of hydropower plants to provide ancillary services, such as frequency regulation and voltage support, which are essential for maintaining grid reliability.

Hydropower's role in energy grids extends beyond electricity generation to include energy storage. Pumped storage hydropower is a form of energy storage that uses excess electricity to pump water from a lower reservoir to an upper reservoir. During periods of high electricity demand, the stored water is released back to the lower reservoir, generating electricity in the process. This ability to store and release energy on demand makes pumped storage hydropower a valuable asset for grid management, providing a buffer against fluctuations in electricity supply and demand.

The environmental benefits of hydropower are significant, as it generates electricity without emitting greenhouse gases or air pollutants. This characteristic makes hydropower a vital tool in the fight against climate change, contributing to the reduction of carbon emissions and supporting the transition to a low-carbon energy system. By displacing fossil fuel-based power generation, hydropower helps to improve air quality and reduce the environmental impact of energy production.

Despite its advantages, the integration of hydropower into energy grids is not without challenges. The construction and operation of hydropower plants can have significant environmental and social impacts, including habitat disruption, changes in water flow, and displacement of communities. These impacts must be carefully managed through comprehensive planning, environmental assessments, and stakeholder engagement to ensure that hydropower projects are sustainable and socially responsible.

Technological advancements are driving innovation in hydropower, enhancing its efficiency and sustainability. Improvements in turbine design and materials are increasing the efficiency of energy conversion, maximizing electricity generation from available water resources. Digital technologies, such as sensors and data analytics, are being used to optimize the operation and maintenance of hydropower plants, reducing costs and improving performance.

The role of hydropower in energy grids is also influenced by policy and regulatory frameworks. Supportive policies, such as renewable energy targets, incentives for sustainable practices, and streamlined permitting processes, create a conducive environment for hydropower development. Policymakers must balance the need for renewable energy expansion with the protection of environmental and social values, ensuring that hydropower projects are both economically viable and ecologically sound.

The future of hydropower in energy grids is shaped by a range of factors, including technological advancements, environmental considerations, and market dynamics. As the global demand for clean and sustainable energy continues to grow, hydropower will play an increasingly important role in meeting this demand. By balancing the benefits and challenges of hydropower, stakeholders can develop strategies that maximize its potential and contribute to a sustainable energy future.

The integration of hydropower into energy grids requires a holistic and adaptive approach, taking into account the complex interactions between energy systems, natural resources, and societal needs. By understanding the unique characteristics and potential of hydropower, stakeholders can unlock its full potential and create a cleaner, more resilient world. Collaboration, innovation, and adaptive management are key to achieving a sustainable energy future that respects and preserves the natural world.

The journey of hydropower is a testament to the power of human ingenuity and the potential of renewable energy. By harnessing the energy of flowing water, we can address the urgent challenges of climate change and energy security while promoting economic development and environmental sustainability. The potential of hydropower is vast, and with continued innovation and collaboration, we can unlock its full potential and create a cleaner, more resilient world for generations to come.

Future Trends and Innovations

The landscape of renewable energy is ever-evolving, driven by technological advancements, shifting market dynamics, and the urgent need to address climate change. As we look to the future, understanding the trends and innovations shaping the renewable energy sector is crucial for stakeholders seeking to harness its full potential. These developments promise to enhance efficiency, sustainability, and accessibility, paving the way for a cleaner and more resilient energy future.

One of the most significant trends in renewable energy is the rapid advancement of technology, which is transforming the way we generate, store, and distribute power. In the realm of solar energy, innovations such as perovskite solar cells are garnering attention for their potential to surpass traditional silicon-based cells in efficiency and cost-effectiveness. These lightweight and flexible materials offer the promise of higher energy conversion rates and lower production costs, making solar power more accessible and affordable.

Wind energy is also experiencing a technological revolution, with the development of larger and more efficient turbines. Offshore wind farms, in particular, are benefiting from advancements in turbine design and materials, allowing them to harness stronger and more consistent winds at sea. Floating wind turbines are emerging as a game-changer, enabling the deployment of wind farms in deeper waters where traditional fixed-bottom turbines are not feasible. These innovations are expanding the geographic reach of wind energy and increasing its contribution to the global energy mix.

Energy storage is another critical area of innovation, as it addresses the intermittent nature of renewable energy sources like wind and solar. Advances in battery technology, particularly in lithium-ion and solid-state batteries, are enhancing the capacity, efficiency, and lifespan of energy storage systems. These improvements are crucial for balancing supply and demand, ensuring a stable and reliable power supply even when the sun isn't shining or the wind isn't blowing. Additionally, the development of grid-scale storage solutions, such as pumped hydro and compressed air energy storage, is providing new avenues for integrating renewable energy into existing power systems.

The rise of digital technologies is revolutionizing the renewable energy sector, enabling smarter and more efficient energy management. The integration of the Internet of Things (IoT), artificial intelligence, and big data analytics is facilitating real-time monitoring and optimization of energy systems. These technologies allow for predictive maintenance, demand forecasting, and grid management, enhancing the reliability and efficiency of renewable energy infrastructure. By leveraging digital tools, stakeholders can optimize energy production, reduce operational costs, and improve the overall performance of renewable energy systems.

Decentralization is a growing trend in the energy sector, driven by the increasing adoption of distributed energy resources (DERs) such as rooftop solar panels, small wind turbines, and microgrids. This shift towards decentralized energy systems empowers consumers to generate and manage their own electricity, reducing reliance on centralized power plants and enhancing energy resilience. Microgrids, in particular, are gaining traction as a solution for remote or underserved communities, providing reliable and sustainable power independent of the main grid. The proliferation of DERs is reshaping the energy landscape, fostering greater energy independence and democratization.

The transition to a low-carbon energy system is also being supported by policy and regulatory frameworks that incentivize renewable energy development. Governments worldwide are setting ambitious renewable energy targets, implementing carbon pricing mechanisms, and offering financial incentives to encourage investment in clean energy technologies. These policies are creating a favorable environment for innovation and growth, driving the expansion of renewable energy capacity and accelerating the shift away from fossil fuels.

Collaboration and partnerships are playing a pivotal role in advancing renewable energy innovations. Cross-sector collaboration between governments, industry, academia, and non-governmental organizations is fostering the exchange of knowledge, resources, and expertise. These partnerships are facilitating research and development, scaling up new technologies, and overcoming barriers to deployment. By working together, stakeholders can accelerate the pace of innovation and achieve shared goals for a sustainable energy future.

The future of renewable energy is also being shaped by changing consumer preferences and societal values. As awareness of environmental issues grows, consumers are increasingly demanding cleaner and more sustainable energy options. This shift in consumer behavior is driving companies to adopt renewable energy solutions and prioritize sustainability in their operations. The rise of corporate renewable energy procurement, through mechanisms such as power purchase agreements (PPAs), is a testament to the growing demand for green energy and the role of businesses in driving the energy transition.

The potential of renewable energy is vast, and the innovations and trends shaping its future are unlocking new opportunities for growth and development. By embracing these advancements, stakeholders can enhance the efficiency, sustainability, and accessibility of renewable energy, contributing to a cleaner and more resilient world. The journey towards a sustainable energy future is complex, but with continued innovation and collaboration, we can overcome the challenges and realize the full potential of renewable energy for generations to come.

Chapter 4: Biomass and Bioenergy: Organic Solutions

Types of Biomass and Conversion Processes

Biomass, a versatile and renewable energy source, has been utilized by humans for millennia. It encompasses a wide range of organic materials, including plant and animal matter, that can be converted into energy. Understanding the various types of biomass and the processes used to convert them into usable energy is essential for harnessing their full potential and integrating them into modern energy systems.

Biomass can be broadly categorized into several types, each with unique characteristics and applications. The most common types include wood and agricultural residues, dedicated energy crops, and organic waste materials. Wood and agricultural residues, such as sawdust, straw, and corn stover, are byproducts of forestry and farming activities. These materials are abundant and readily available, making them a popular choice for biomass energy production.

Dedicated energy crops, such as switchgrass, miscanthus, and willow, are specifically cultivated for energy purposes. These crops are chosen for their high yield and low input requirements, making them an efficient and sustainable source of biomass. They can be grown on marginal lands, reducing competition with food crops and minimizing environmental impacts.

Organic waste materials, including municipal solid waste, animal manure, and food waste, offer another valuable source of biomass. These materials are often considered a burden, but they can be transformed into energy through various conversion processes, providing a sustainable solution to waste management challenges.

The conversion of biomass into energy involves several processes, each suited to different types of biomass and end-use applications. The primary conversion processes include combustion, gasification, pyrolysis, anaerobic digestion, and fermentation.

Combustion is the most straightforward and widely used method for converting biomass into energy. It involves burning biomass in the presence of oxygen to produce heat, which can be used directly for heating or to generate electricity through steam turbines. Combustion is well-suited for woody biomass and agricultural residues, providing a reliable and efficient energy source. However, it requires careful management to minimize emissions and ensure environmental sustainability.

Gasification is a more advanced conversion process that involves heating biomass in a low-oxygen environment to produce a combustible gas known as syngas. Syngas, composed primarily of carbon monoxide and hydrogen, can be used to generate electricity, produce heat, or serve as a feedstock for chemical synthesis. Gasification offers higher efficiency and lower emissions compared to combustion, making it an attractive option for biomass conversion.

Pyrolysis is a thermal decomposition process that occurs in the absence of oxygen, breaking down biomass into bio-oil, syngas, and biochar. Bio-oil can be refined into liquid fuels, while syngas can be used for energy production. Biochar, a carbon-rich solid, can be used as a soil amendment to improve soil fertility and sequester carbon. Pyrolysis is a versatile process that can be applied to a wide range of biomass types, offering multiple products and environmental benefits.

Anaerobic digestion is a biological process that converts organic waste materials into biogas through the action of microorganisms in an oxygen-free environment. Biogas, composed mainly of methane and carbon dioxide, can be used for heating, electricity generation, or as a vehicle fuel. Anaerobic digestion is particularly well-suited for wet biomass, such as animal manure and food waste, providing a sustainable solution for waste management and energy production.

Fermentation is a biochemical process that converts sugars in biomass into ethanol, a renewable liquid fuel. This process is commonly used for producing biofuels from sugarcane, corn, and other carbohydrate-rich crops. Ethanol can be blended with gasoline to reduce emissions and enhance energy security. Advances in fermentation technology are enabling the use of lignocellulosic biomass, such as agricultural residues and dedicated energy crops, for ethanol production, expanding the range of feedstocks and improving sustainability.

The choice of conversion process depends on several factors, including the type of biomass, desired end products, and economic considerations. Each process has its advantages and limitations, and selecting the most appropriate method requires a comprehensive understanding of the specific context and objectives.

The integration of biomass energy into modern energy systems offers numerous benefits, including reduced greenhouse gas emissions, enhanced energy security, and sustainable waste management. By utilizing locally available biomass resources, communities can reduce their reliance on fossil fuels and promote energy independence. Additionally, the use of biomass for energy production can create economic opportunities in rural areas, supporting local agriculture and forestry industries.

However, the development of biomass energy systems must be carefully managed to ensure environmental sustainability and social acceptance. The cultivation of dedicated energy crops should avoid competition with food production and minimize impacts on biodiversity and water resources. The use of organic waste materials for energy production should prioritize waste reduction and recycling, ensuring that energy recovery is part of a comprehensive waste management strategy.

Technological advancements and innovation are driving the evolution of biomass energy, enhancing efficiency and sustainability. Improvements in conversion technologies, such as advanced gasification and pyrolysis systems, are increasing the efficiency and reducing the environmental impact of biomass energy production. The development of integrated biorefineries, which produce multiple products from biomass, is expanding the range of applications and improving the economic viability of biomass energy systems.

The future of biomass energy is shaped by a range of factors, including technological advancements, policy frameworks, and market dynamics. Supportive policies, such as renewable energy targets, incentives for sustainable practices, and research funding, are essential for promoting the development and deployment of biomass energy technologies. Collaboration and partnerships between governments, industry, and research institutions are crucial for overcoming barriers and accelerating the transition to a sustainable energy future.

By understanding the types of biomass and conversion processes, stakeholders can harness the full potential of this versatile energy source and contribute to a cleaner, more resilient world. The journey towards a sustainable energy future is complex, but with continued innovation and collaboration, we can overcome the challenges and realize the full potential of biomass energy for generations to come.

Biofuels: From Production to Consumption

Biofuels have emerged as a promising alternative to fossil fuels, offering a renewable and potentially carbon-neutral source of energy. Derived from organic materials, biofuels can be used in transportation, heating, and electricity generation, providing a versatile solution to the growing demand for sustainable energy. Understanding the journey of biofuels from production to consumption is essential for maximizing their benefits and addressing the challenges they present.

The production of biofuels begins with the cultivation or collection of biomass feedstocks. These feedstocks can be broadly categorized into three main types: first-generation, second-generation, and third-generation. First-generation biofuels are produced from food crops such as corn, sugarcane, and soybeans. These crops are rich in sugars, starches, or oils, which can be easily converted into biofuels through fermentation or transesterification processes. While first-generation biofuels are well-established and widely used, they have faced criticism for competing with food production and contributing to land-use changes.

Second-generation biofuels address some of the limitations of their predecessors by utilizing non-food biomass, such as agricultural residues, forestry byproducts, and dedicated energy crops like switchgrass and miscanthus. These feedstocks are often lignocellulosic, meaning they contain complex carbohydrates that require advanced conversion technologies to break down into fermentable sugars. The development of second-generation biofuels is driven by the need to reduce competition with food crops and improve the sustainability of biofuel production.

Third-generation biofuels represent the cutting edge of biofuel technology, utilizing algae and other microorganisms as feedstocks. Algae have the potential to produce high yields of oils and carbohydrates, which can be converted into biodiesel and bioethanol, respectively. The cultivation of algae can occur in a variety of environments, including wastewater and non-arable land, minimizing competition with traditional agriculture. Additionally, algae can capture carbon dioxide during growth, offering the potential for carbon-neutral or even carbon-negative biofuel production.

Once the biomass feedstocks are harvested or collected, they undergo a series of conversion processes to produce biofuels. The most common biofuels include bioethanol, biodiesel, and biogas. Bioethanol is produced through the fermentation of sugars and starches by microorganisms, resulting in a liquid fuel that can be blended with gasoline or used in flex-fuel vehicles. Biodiesel is produced through the transesterification of oils and fats, creating a renewable diesel substitute that can be used in diesel engines with little or no modification. Biogas is generated through the anaerobic digestion of organic matter, producing a mixture of methane and carbon dioxide that can be used for heating, electricity generation, or as a vehicle fuel.

The conversion of biomass into biofuels involves several key technologies and processes. For bioethanol production, the primary steps include pretreatment, hydrolysis, fermentation, and distillation. Pretreatment involves breaking down the complex structure of lignocellulosic biomass to make it more accessible to enzymes. Hydrolysis uses enzymes to convert cellulose and hemicellulose into simple sugars, which are then fermented by microorganisms to produce ethanol. Distillation is used to purify the ethanol, removing water and other impurities to achieve the desired concentration.

Biodiesel production involves the extraction of oils from biomass feedstocks, followed by transesterification. This process involves reacting the oils with an alcohol, typically methanol, in the presence of a catalyst to produce biodiesel and glycerin. The biodiesel is then purified to remove any remaining impurities, ensuring it meets quality standards for use in diesel engines.

Biogas production relies on anaerobic digestion, a biological process that occurs in the absence of oxygen. Organic matter is broken down by microorganisms, producing biogas and digestate. The biogas is collected and purified to remove impurities, such as hydrogen sulfide and moisture, before being used as a fuel. The digestate, a nutrient-rich byproduct, can be used as a fertilizer or soil amendment, closing the nutrient loop and enhancing the sustainability of the process.

The consumption of biofuels is influenced by a range of factors, including availability, cost, and compatibility with existing infrastructure. Bioethanol is commonly used as a gasoline additive, with blends such as E10 (10% ethanol, 90% gasoline) and E85 (85% ethanol, 15% gasoline) widely available in many countries. Biodiesel is used in diesel engines, either as a pure fuel (B100) or blended with petroleum diesel (e.g., B20, 20% biodiesel, 80% diesel). Biogas can be used for heating, electricity generation, or as a vehicle fuel, often in the form of compressed natural gas (CNG) or liquefied natural gas (LNG).

The integration of biofuels into the energy landscape offers numerous benefits, including reduced greenhouse gas emissions, enhanced energy security, and support for rural economies. By displacing fossil fuels, biofuels contribute to the reduction of carbon emissions and help mitigate the impacts of climate change. The use of locally sourced biomass feedstocks reduces reliance on imported fuels, enhancing energy independence and resilience. Additionally, the production of biofuels can create economic opportunities in rural areas, supporting agriculture and forestry industries.

However, the development and consumption of biofuels must be carefully managed to ensure environmental sustainability and social acceptance. The cultivation of biomass feedstocks should avoid competition with food production and minimize impacts on biodiversity and water resources. The conversion processes should prioritize efficiency and minimize emissions, ensuring that biofuels deliver genuine environmental benefits.

Technological advancements and innovation are driving the evolution of biofuels, enhancing their efficiency and sustainability. Improvements in conversion technologies, such as advanced fermentation and transesterification processes, are increasing yields and reducing costs. The development of integrated biorefineries, which produce multiple products from biomass, is expanding the range of applications and improving the economic viability of biofuel production.

The future of biofuels is shaped by a range of factors, including technological advancements, policy frameworks, and market dynamics. Supportive policies, such as renewable fuel standards, incentives for sustainable practices, and research funding, are essential for promoting the development and deployment of biofuel technologies. Collaboration and partnerships between governments, industry, and research institutions are crucial for overcoming barriers and accelerating the transition to a sustainable energy future.

By understanding the journey of biofuels from production to consumption, stakeholders can harness the full potential of this renewable energy source and contribute to a cleaner, more resilient world. The journey towards a sustainable energy future is complex, but with continued innovation and collaboration, we can overcome the challenges and realize the full potential of biofuels for generations to come.

Environmental and Economic Implications

The intersection of environmental and economic considerations is a critical aspect of the renewable energy landscape. As the world grapples with the dual challenges of climate change and economic development, understanding the implications of renewable energy adoption is essential for crafting policies and strategies that balance ecological sustainability with economic growth.

Renewable energy sources, such as solar, wind, hydropower, and bioenergy, offer significant environmental benefits compared to fossil fuels. They produce little to no greenhouse gas emissions during operation, contributing to the reduction of carbon footprints and helping mitigate the impacts of climate change. By displacing fossil fuel-based power generation, renewables improve air quality, reduce acid rain, and decrease the incidence of respiratory and cardiovascular diseases linked to air pollution. These environmental benefits translate into substantial public health savings and enhanced quality of life.

The transition to renewable energy also has profound implications for biodiversity and ecosystems. Unlike fossil fuel extraction, which often leads to habitat destruction and pollution, renewable energy projects can be designed to minimize ecological impacts. For instance, solar farms can be integrated with agricultural activities, allowing for dual land use, while wind farms can be sited to avoid migratory bird paths. Hydropower projects, when carefully managed, can maintain riverine ecosystems and support aquatic biodiversity. By prioritizing sustainable practices, renewable energy development can coexist with conservation efforts, preserving natural habitats and promoting biodiversity.

Economically, the shift to renewable energy presents both opportunities and challenges. On one hand, the renewable energy sector is a significant driver of job creation and economic growth. The International Renewable Energy Agency (IRENA) estimates that the renewable energy industry employed over 11 million people worldwide in recent years, with the potential for millions more as the sector expands. Jobs in renewable energy span a wide range of roles, from manufacturing and installation to operations and maintenance, offering opportunities for skilled and unskilled workers alike. This job creation potential is particularly valuable in regions facing high unemployment or economic stagnation.

Renewable energy projects also stimulate local economies by attracting investment, increasing tax revenues, and supporting ancillary industries. For example, the construction of a wind farm can boost demand for local services, such as transportation, hospitality, and construction materials. Additionally, renewable energy can enhance energy security by reducing dependence on imported fuels, stabilizing energy prices, and insulating economies from volatile fossil fuel markets. This energy independence is particularly important for countries with limited domestic fossil fuel resources, allowing them to redirect funds towards other critical areas of development.

However, the economic implications of renewable energy adoption are not uniformly positive. The transition from fossil fuels to renewables can lead to job losses in traditional energy sectors, such as coal mining and oil extraction. These industries often form the backbone of local economies, and their decline can have significant social and economic repercussions. To address these challenges, policymakers must implement strategies for a just transition, providing support for affected workers and communities through retraining programs, economic diversification, and social safety nets.

The cost of renewable energy technologies has been a significant barrier to widespread adoption, but recent years have seen dramatic declines in costs, making renewables increasingly competitive with fossil fuels. The cost of solar photovoltaic (PV) systems, for example, has fallen by over 80%

in the past decade, while wind power costs have decreased by nearly 50%. These cost reductions are driven by technological advancements, economies of scale, and increased competition in the market. As a result, renewables are becoming the most cost-effective option for new power generation in many regions, offering economic benefits alongside environmental gains.

Despite these cost reductions, the initial capital investment required for renewable energy projects can still be a hurdle, particularly for developing countries with limited financial resources. Access to affordable financing is crucial for overcoming this barrier and enabling the deployment of renewable energy technologies. Innovative financing mechanisms, such as green bonds, public-private partnerships, and international climate funds, are playing a vital role in mobilizing capital for renewable energy projects. By leveraging these financial tools, countries can accelerate the transition to a sustainable energy future.

The integration of renewable energy into existing energy systems also presents technical and economic challenges. The intermittent nature of some renewable sources, such as solar and wind, requires the development of energy storage solutions and grid management technologies to ensure a stable and reliable power supply. Investments in grid infrastructure, smart grid technologies, and energy storage systems are essential for accommodating the growing share of renewables in the energy mix. These investments, while costly, offer long-term economic benefits by enhancing grid resilience, reducing transmission losses, and enabling more efficient energy use.

The environmental and economic implications of renewable energy adoption are closely intertwined, and achieving a balance between the two requires careful planning and collaboration among stakeholders. Policymakers, industry leaders, and communities must work together to develop strategies that maximize the benefits of renewable energy while addressing the associated challenges. This collaborative approach is essential for ensuring that the transition to renewable energy is both environmentally sustainable and economically inclusive.

The journey towards a sustainable energy future is complex, but the potential rewards are immense. By embracing renewable energy, we can reduce our environmental impact, enhance energy security, and drive economic growth. The path forward requires innovation, investment, and a commitment to sustainability, but with the right policies and partnerships, we can create a cleaner, more resilient world for future generations.

Challenges in Scaling Bioenergy

Scaling bioenergy to meet global energy demands presents a complex array of challenges that require innovative solutions and strategic planning. As the world seeks to transition to more sustainable energy sources, bioenergy offers a promising alternative. However, its expansion is not without obstacles, ranging from technological and economic hurdles to environmental and social considerations.

One of the primary challenges in scaling bioenergy is the availability and sustainability of biomass feedstocks. The competition for land use between bioenergy crops and food production is a significant concern, as it can lead to food security issues and increased food prices. To address this, it is crucial to identify and utilize marginal lands that are unsuitable for food crops, as well as to develop high-yield, low-input energy crops that minimize resource use. Additionally, the use of agricultural residues and waste materials as feedstocks can alleviate pressure on land resources and contribute to a circular economy.

The logistics of biomass collection, transportation, and storage also pose significant challenges. Biomass is often bulky and has a low energy density, making it costly and inefficient to transport over long distances. Developing efficient supply chains and infrastructure is essential for reducing costs and ensuring a reliable supply of feedstocks. Innovations in biomass densification, such as pelletization and briquetting, can improve transport efficiency and storage stability, making bioenergy more economically viable.

Technological advancements are critical for overcoming the challenges associated with bioenergy conversion processes. Many current technologies, such as those used for second-generation biofuels, are still in the early stages of development and require further research and optimization to become commercially viable. Investing in research and development is essential for improving conversion efficiencies, reducing costs, and minimizing environmental impacts. Collaboration between academia, industry, and government can accelerate the development and deployment of advanced bioenergy technologies.

Economic factors play a significant role in the scalability of bioenergy. The initial capital investment required for bioenergy projects can be a barrier, particularly in developing countries with limited financial resources. Access to affordable financing and incentives, such as subsidies and tax credits, are crucial for encouraging investment in bioenergy infrastructure. Additionally, establishing stable and predictable policy frameworks can provide the certainty needed for long-term investments and market development.

The integration of bioenergy into existing energy systems presents technical challenges, particularly in terms of grid compatibility and energy storage. The intermittent nature of some bioenergy sources, such as biogas, requires the development of flexible and responsive grid management systems to ensure a stable and reliable power supply. Investments in smart grid technologies and energy storage solutions are essential for accommodating the growing share of bioenergy in the energy mix and enhancing grid resilience.

Environmental considerations are paramount when scaling bioenergy. While bioenergy has the potential to reduce greenhouse gas emissions compared to fossil fuels, its environmental benefits depend on sustainable feedstock production and efficient conversion processes. Unsustainable practices, such as deforestation for bioenergy crop cultivation, can negate the environmental advantages of bioenergy and lead to biodiversity loss and ecosystem degradation. Implementing sustainable land management practices and certifying bioenergy supply chains can help ensure that bioenergy contributes positively to environmental goals.

Social acceptance and public perception are also critical factors in the successful scaling of bioenergy. Concerns about land use changes, food security, and environmental impacts can lead to opposition from local communities and stakeholders. Engaging with communities and stakeholders through transparent communication and participatory decision-making processes can build trust and support for bioenergy projects. Education and awareness campaigns can also play a role in highlighting the benefits of bioenergy and addressing misconceptions.

The global nature of bioenergy challenges necessitates international cooperation and knowledge sharing. Countries can learn from each other's experiences and best practices, fostering innovation and accelerating the adoption of bioenergy technologies. International organizations and partnerships can facilitate collaboration and provide technical and financial support to countries seeking to develop their bioenergy sectors.

Despite the challenges, the potential benefits of scaling bioenergy are significant. Bioenergy can contribute to energy security, reduce greenhouse gas emissions, and support rural development by creating jobs and economic opportunities. By addressing the challenges and implementing strategic solutions, bioenergy can play a vital role in the transition to a sustainable energy future.

The path forward requires a holistic approach that considers the interconnected nature of the challenges and leverages the strengths of diverse stakeholders. By fostering innovation, collaboration, and sustainable practices, the bioenergy sector can overcome the obstacles to scaling and unlock its full potential as a key component of the global energy landscape.

Innovations and Future Directions

Innovation is the lifeblood of progress, and in the realm of renewable energy, it is the key to unlocking a sustainable future. As the world grapples with the urgent need to transition away from fossil fuels, the role of innovation in shaping the future of energy cannot be overstated. From cutting-edge technologies to novel approaches in policy and infrastructure, the landscape of renewable energy is evolving rapidly, offering exciting possibilities and new directions.

One of the most promising areas of innovation lies in the development of advanced materials and technologies for energy capture and conversion. Solar energy, for instance, has seen remarkable advancements with the introduction of perovskite solar cells. These materials offer the potential for higher efficiency and lower production costs compared to traditional silicon-based cells. Researchers are exploring ways to enhance the stability and scalability of perovskite cells, aiming to bring them to market as a viable alternative for large-scale solar power generation.

Wind energy is also benefiting from technological innovations. The design and engineering of wind turbines have evolved significantly, with the development of larger and more efficient turbines capable of harnessing wind energy at lower speeds. Offshore wind farms, in particular, are gaining traction as a promising avenue for expanding wind energy capacity. Floating wind turbines, which can be deployed in deeper waters, are opening up new possibilities for harnessing wind energy in regions previously considered unsuitable.

Energy storage is another critical area where innovation is driving change. The intermittent nature of renewable energy sources like solar and wind necessitates effective storage solutions to ensure a stable and reliable power supply. Advances in battery technology, such as lithium-sulfur and solid-state batteries, are paving the way for more efficient and longer-lasting energy storage systems. These innovations are crucial for enabling the widespread adoption of electric vehicles and integrating renewable energy into the grid.

Beyond technological advancements, innovation in policy and market mechanisms is essential for accelerating the transition to renewable energy. Governments and policymakers are exploring new approaches to incentivize renewable energy adoption, such as feed-in tariffs, renewable portfolio standards, and carbon pricing. These mechanisms aim to create a favorable environment for investment in renewable energy projects and encourage the development of clean energy technologies.

The concept of decentralized energy systems is gaining momentum as a future direction for renewable energy. Distributed energy resources, such as rooftop solar panels and small-scale wind turbines, allow individuals and communities to generate their own electricity, reducing reliance on centralized power plants. This shift towards decentralized energy systems is supported by advancements in smart grid technology, which enables efficient energy management and distribution at the local level.

The integration of digital technologies, such as artificial intelligence and the Internet of Things (IoT), is transforming the renewable energy sector. These technologies enable real-time monitoring and optimization of energy systems, improving efficiency and reducing operational costs. For example, AI algorithms can predict energy demand and optimize the operation of renewable energy assets, while IoT devices can facilitate demand response and energy management in smart homes and buildings.

Bioenergy, too, is witnessing innovative approaches that promise to enhance its sustainability and scalability. The development of advanced biofuels, such as cellulosic ethanol and algae-based biodiesel, is expanding the range of feedstocks and improving the environmental performance of biofuels. Innovations in biorefinery processes are enabling the production of multiple products from biomass, enhancing the economic viability of bioenergy projects.

The circular economy concept is gaining traction as a framework for sustainable resource management in the renewable energy sector. By designing systems that minimize waste and maximize resource efficiency, the circular economy aims to create closed-loop systems where materials are reused and recycled. This approach is particularly relevant for bioenergy, where the use of agricultural residues and waste materials as feedstocks can contribute to a more sustainable and resilient energy system.

Collaboration and partnerships are essential for driving innovation and shaping the future of renewable energy. Cross-sector collaboration between governments, industry, academia, and civil society can facilitate knowledge sharing, accelerate technological development, and overcome barriers to adoption. International cooperation is also crucial for addressing global challenges and ensuring that the benefits of renewable energy are accessible to all.

Education and capacity building play a vital role in preparing the workforce for the renewable energy transition. As the sector grows, there is a need for skilled professionals who can design, implement, and maintain renewable energy systems. Educational institutions and training programs are evolving to meet this demand, offering courses and certifications in renewable energy technologies and sustainable practices.

Public engagement and awareness are key to fostering support for renewable energy initiatives. By involving communities in the planning and decision-making processes, stakeholders can build trust and ensure that projects align with local needs and values. Public awareness campaigns can also highlight the benefits of renewable energy and encourage individuals to adopt sustainable practices in their daily lives.

The future of renewable energy is bright, but it requires a concerted effort to overcome the challenges and seize the opportunities that lie ahead. By embracing innovation and exploring new directions, we can accelerate the transition to a sustainable energy future and create a cleaner, more resilient world for generations to come. The journey is complex, but with determination and collaboration, the potential for positive change is limitless.

Chapter 5: Geothermal Energy: Tapping Earth's Heat

Geothermal Systems and Technologies

Geothermal energy, a powerful yet often underutilized resource, taps into the Earth's internal heat to provide a sustainable and reliable energy source. This form of energy has been harnessed for centuries, from ancient Roman baths to modern power plants, and its potential continues to grow with advancements in technology and exploration. Understanding the intricacies of geothermal systems and technologies is essential for unlocking their full potential and integrating them into the broader energy landscape.

At the heart of geothermal energy lies the Earth's natural heat, which originates from the planet's formation and the radioactive decay of minerals. This heat is stored in rocks and fluids beneath the Earth's crust and can be accessed through various geothermal systems. These systems are broadly categorized into three main types: hydrothermal, enhanced geothermal systems (EGS), and direct-use applications.

Hydrothermal systems are the most common and commercially developed form of geothermal energy. They rely on naturally occurring reservoirs of hot water and steam, typically found in regions with high volcanic activity or tectonic plate boundaries. These reservoirs are accessed by drilling wells, similar to those used in the oil and gas industry, to bring the hot fluids to the surface. Once extracted, the steam or hot water is used to drive turbines connected to electricity generators, producing clean and renewable power. The cooled water is then reinjected into the reservoir to sustain the resource and maintain pressure.

Enhanced geothermal systems (EGS) represent a more advanced and versatile approach to geothermal energy. Unlike hydrothermal systems, EGS do not rely on naturally occurring reservoirs. Instead, they create artificial reservoirs by injecting water into hot, dry rock formations deep beneath the Earth's surface. This process, known as hydraulic stimulation, fractures the rock and creates pathways for the water to circulate and absorb heat. The heated water is then pumped back to the surface, where it can be used for electricity generation or direct heating applications. EGS have the potential to significantly expand the geographic availability of geothermal energy, making it accessible in regions without natural hydrothermal resources.

Direct-use applications of geothermal energy involve the use of geothermal heat for non-electric purposes, such as space heating, greenhouse agriculture, aquaculture, and industrial processes. These applications typically utilize lower-temperature geothermal resources, which are more widely available than high-temperature resources required for electricity generation. Direct-use systems often involve the use of heat exchangers to transfer geothermal heat to a secondary fluid, which is then circulated through buildings or facilities to provide heating.

The development and deployment of geothermal technologies face several challenges, including high upfront costs, resource exploration risks, and environmental considerations. Drilling and exploration account for a significant portion of the costs associated with geothermal projects, as they require specialized equipment and expertise. The success of a geothermal project depends on accurately locating and characterizing geothermal resources, which can be a complex and uncertain process.

Advances in geophysical and geochemical exploration techniques, such as seismic imaging and remote sensing, are helping to reduce these risks and improve the success rate of geothermal projects.

Environmental considerations are also an important aspect of geothermal energy development. While geothermal energy is generally considered environmentally friendly, it can have localized impacts, such as land subsidence, induced seismicity, and the release of greenhouse gases and other pollutants from geothermal fluids. Proper site selection, monitoring, and management practices are essential for minimizing these impacts and ensuring the sustainability of geothermal resources.

The integration of geothermal energy into existing energy systems offers numerous benefits, including baseload power generation, grid stability, and reduced greenhouse gas emissions. Unlike solar and wind energy, which are intermittent and weather-dependent, geothermal energy provides a constant and reliable power supply, making it an ideal complement to other renewable energy sources. By providing baseload power, geothermal energy can enhance grid stability and reduce the need for fossil fuel-based backup power.

The future of geothermal energy is bright, with ongoing research and development efforts aimed at overcoming existing challenges and unlocking new opportunities. Innovations in drilling technology, such as the use of advanced materials and techniques, are reducing costs and improving the efficiency of geothermal projects. The development of hybrid systems, which combine geothermal energy with other renewable sources, is expanding the range of applications and enhancing the overall efficiency of energy systems.

Geothermal energy also holds promise for contributing to the decarbonization of heating and cooling sectors, which account for a significant portion of global energy consumption and greenhouse gas emissions. Ground-source heat pumps, which utilize the stable temperatures of the Earth's subsurface to provide heating and cooling, are an increasingly popular and efficient option for residential and commercial buildings. By reducing reliance on fossil fuels for heating and cooling, geothermal technologies can play a crucial role in achieving climate goals and transitioning to a sustainable energy future.

International collaboration and knowledge sharing are essential for advancing geothermal energy and realizing its full potential. Countries with extensive geothermal experience, such as Iceland, New Zealand, and the United States, can provide valuable insights and expertise to emerging geothermal markets. International organizations and partnerships can facilitate the exchange of best practices, technical assistance, and financial support, helping to accelerate the development and deployment of geothermal technologies worldwide.

Education and public awareness are also critical components of geothermal energy advancement. By increasing understanding of geothermal energy and its benefits, stakeholders can build support for geothermal projects and encourage investment in research and development. Educational programs and outreach initiatives can help train the next generation of geothermal professionals and foster a culture of innovation and sustainability.

The journey towards a sustainable energy future is complex and multifaceted, but geothermal energy offers a powerful and versatile tool for achieving this goal. By embracing innovation, collaboration, and sustainable practices, the geothermal sector can overcome existing challenges and unlock new opportunities, paving the way for a cleaner, more resilient world. The potential for positive change is immense, and with determination and cooperation, geothermal energy can play a pivotal role in shaping the future of energy.

Global Geothermal Hotspots

The Earth's crust is a dynamic tapestry of geothermal activity, with certain regions standing out as exceptional sources of geothermal energy. These global geothermal hotspots are characterized by their unique geological features, which create ideal conditions for harnessing the Earth's internal heat. Understanding these hotspots provides valuable insights into the potential of geothermal energy and the diverse ways it can be utilized.

One of the most renowned geothermal hotspots is Iceland, a country that sits atop the Mid-Atlantic Ridge, where the North American and Eurasian tectonic plates diverge. This geological setting results in abundant geothermal resources, which Iceland has harnessed to become a global leader in renewable energy. Over 90% of Iceland's homes are heated with geothermal energy, and the country generates a significant portion of its electricity from geothermal power plants. The Blue Lagoon, a famous geothermal spa, exemplifies how Iceland has integrated geothermal energy into its tourism industry, showcasing the versatility of this resource.

The Pacific Ring of Fire, a horseshoe-shaped area encircling the Pacific Ocean, is another prominent geothermal hotspot. This region is characterized by intense volcanic and seismic activity, providing ample opportunities for geothermal energy development. Countries like the Philippines, Indonesia, and New Zealand have capitalized on their location within the Ring of Fire to develop robust geothermal industries. The Philippines, for instance, ranks among the world's top producers of geothermal energy, with numerous power plants tapping into the country's volcanic resources. Indonesia, with its vast geothermal potential, is actively expanding its geothermal capacity to meet growing energy demands and reduce reliance on fossil fuels.

In the United States, the western states are home to significant geothermal resources, particularly in California and Nevada. The Geysers, located in Northern California, is the largest geothermal field in the world, with a capacity of over 1,500 megawatts. This field has been a cornerstone of California's renewable energy portfolio for decades, providing a reliable source of baseload power. Nevada, known for its geothermal-friendly geology, has also developed a thriving geothermal industry, with numerous power plants and direct-use applications contributing to the state's energy mix.

East Africa's Rift Valley is another notable geothermal hotspot, with countries like Kenya leading the way in geothermal development. The Rift Valley's unique geological features, including active volcanoes and tectonic activity, create ideal conditions for geothermal energy. Kenya's Olkaria Geothermal

Plant is one of the largest in Africa, providing a significant portion of the country's electricity and supporting economic growth. The success of geothermal projects in Kenya has inspired neighboring countries to explore their geothermal potential, positioning East Africa as a burgeoning hub for geothermal energy.

Japan, a country with a long history of utilizing geothermal resources for hot springs and bathing, is increasingly turning to geothermal energy as a means of diversifying its energy sources. The country's mountainous terrain and volcanic activity offer significant geothermal potential, which is being tapped to reduce dependence on imported fossil fuels and nuclear power. Japan's geothermal industry is characterized by a blend of traditional and modern applications, with hot spring resorts coexisting alongside cutting-edge geothermal power plants.

Italy, home to the world's first geothermal power plant in Larderello, continues to be a pioneer in geothermal energy. The Larderello field, located in Tuscany, has been producing geothermal electricity since the early 20th century and remains a key component of Italy's renewable energy strategy. The country's geothermal resources are not limited to electricity generation; they also support a range of direct-use applications, including district heating and industrial processes.

Turkey, situated on the seismically active Anatolian Plate, has emerged as a significant player in the geothermal sector. The country's geothermal resources are concentrated in the western regions, where high-temperature reservoirs provide opportunities for electricity generation and direct-use applications. Turkey's geothermal industry has experienced rapid growth in recent years, driven by supportive government policies and investment in exploration and development.

Central America's volcanic landscape offers abundant geothermal resources, with countries like Costa Rica and El Salvador leading the way in geothermal energy development. Costa Rica, known for its commitment to renewable energy, has integrated geothermal power into its energy mix, contributing to the country's goal of achieving carbon neutrality. El Salvador, with its geothermal-rich terrain, has developed several power plants that provide a substantial portion of the country's electricity, reducing reliance on imported fossil fuels.

The potential of geothermal energy extends beyond these hotspots, with many regions around the world exploring their geothermal resources. Countries in Europe, such as Germany and France, are investing in geothermal projects to diversify their energy sources and reduce greenhouse gas emissions. In Asia, countries like China and South Korea are exploring geothermal energy as part of their broader renewable energy strategies.

The development of geothermal energy in these global hotspots offers valuable lessons and best practices for other regions seeking to harness their geothermal potential. Key factors for success include supportive policy frameworks, investment in research and development, and collaboration between government, industry, and academia. By sharing knowledge and expertise, countries can accelerate the deployment of geothermal technologies and contribute to a sustainable energy future.

Geothermal energy's ability to provide baseload power, reduce greenhouse gas emissions, and support economic development makes it a vital component of the global energy transition. As the world seeks to address the challenges of climate change and energy security, the exploration and development of geothermal hotspots offer a promising path forward. By embracing innovation and collaboration, the geothermal sector can unlock new opportunities and play a pivotal role in shaping the future of energy.

Environmental and Economic Benefits

Harnessing renewable energy sources offers a multitude of environmental and economic benefits that are crucial for a sustainable future. As the world grapples with the impacts of climate change and the depletion of fossil fuels, the transition to renewable energy presents a viable solution to these pressing challenges. By understanding the advantages of renewable energy, individuals, businesses, and governments can make informed decisions that contribute to a cleaner and more resilient world.

One of the most significant environmental benefits of renewable energy is the reduction of greenhouse gas emissions. Unlike fossil fuels, which release carbon dioxide and other pollutants when burned, renewable energy sources such as solar, wind, and geothermal produce little to no emissions during operation. This reduction in emissions is essential for mitigating climate change and limiting global temperature rise. By replacing fossil fuel-based power generation with renewable energy, countries can significantly decrease their carbon footprint and contribute to international climate goals.

Renewable energy also plays a vital role in reducing air and water pollution. Traditional energy sources, such as coal and natural gas, release harmful pollutants into the air, including sulfur dioxide, nitrogen oxides, and particulate matter. These pollutants contribute to smog, acid rain, and respiratory illnesses, posing a threat to human health and the environment. In contrast, renewable energy sources generate electricity without emitting these harmful substances, leading to cleaner air and improved public health outcomes. Additionally, renewable energy technologies typically require less water for operation compared to conventional power plants, reducing the strain on water resources and minimizing the risk of water pollution.

The preservation of natural habitats and biodiversity is another important environmental benefit of renewable energy. Fossil fuel extraction and combustion can lead to habitat destruction, deforestation, and the loss of biodiversity. Renewable energy projects, when carefully planned and implemented, can minimize these impacts by utilizing existing infrastructure, such as rooftops for solar panels, or by siting projects on degraded or marginal lands. By reducing the need for resource extraction and minimizing land disturbance, renewable energy can help protect ecosystems and preserve biodiversity.

On the economic front, renewable energy offers numerous advantages that contribute to economic growth and job creation. The renewable energy sector is one of the fastest-growing industries globally, driven by technological advancements, decreasing costs, and supportive policies. As the demand for clean energy continues to rise, the sector is creating millions of jobs in manufacturing, installation, maintenance, and research and development. These jobs are often more sustainable and resilient to economic fluctuations compared to those in the fossil fuel industry, providing stable employment opportunities for communities around the world.

The decentralization of energy production is another economic benefit of renewable energy. Distributed energy resources, such as rooftop solar panels and small-scale wind turbines, allow individuals and communities to generate their own electricity, reducing reliance on centralized power plants and enhancing energy security. This decentralization can lead to cost savings for consumers, as they can offset their energy bills and potentially sell excess electricity back to the grid. Moreover, decentralized energy systems can increase resilience to natural disasters and grid disruptions, providing a reliable power supply in times of need.

Renewable energy also offers the potential for energy independence and security. By diversifying energy sources and reducing reliance on imported fossil fuels, countries can enhance their energy security and reduce vulnerability to geopolitical tensions and price volatility. This energy independence can lead to greater economic stability and resilience, as countries are less exposed to the risks associated with fluctuating fossil fuel markets.

The economic benefits of renewable energy extend to rural and remote communities, where access to reliable and affordable energy can be limited. Renewable energy projects, such as wind farms and solar installations, can provide much-needed infrastructure and investment in these areas, supporting local economies and improving quality of life. By creating jobs and generating revenue, renewable energy can contribute to rural development and reduce poverty, fostering social and economic well-being.

The decreasing costs of renewable energy technologies are a key driver of their economic benefits. Over the past decade, the cost of solar and wind energy has plummeted, making them increasingly competitive with traditional energy sources. This cost reduction is largely due to economies of scale, technological advancements, and increased competition in the market. As renewable energy becomes more affordable, it is becoming an attractive option for businesses and consumers, leading to increased adoption and investment.

The integration of renewable energy into existing energy systems can also lead to cost savings and efficiency improvements. By reducing the need for fossil fuel-based power generation, renewable energy can lower fuel costs and decrease the wear and tear on conventional power plants. Additionally, the use of smart grid technologies and energy storage solutions can optimize the operation of renewable energy systems, enhancing their efficiency and reliability.

The environmental and economic benefits of renewable energy are interconnected and mutually reinforcing. By reducing emissions and pollution, renewable energy contributes to a healthier environment, which in turn supports economic growth and development. Similarly, the economic advantages of renewable energy, such as job creation and cost savings, can drive further investment in clean energy technologies, accelerating the transition to a sustainable energy future.

The path to a sustainable future requires a concerted effort to overcome the challenges and seize the opportunities presented by renewable energy. By embracing innovation, collaboration, and sustainable practices, individuals, businesses, and governments can unlock the full potential of renewable energy and create a cleaner, more resilient world. The journey is complex, but with determination and cooperation, the potential for positive change is immense.

Printed in the USA
CPSIA information can be obtained
at www.ICGtesting.com
CBHW072000291124
18173CB00045B/862